# PICTURING CLASS

# PICTURING CLASS

## LEWIS W. HINE PHOTOGRAPHS
## CHILD LABOR IN NEW ENGLAND

———

## Robert Macieski

University of Massachusetts Press  *Amherst and Boston*

Copyright © 2015 by University of Massachusetts Press
All rights reserved
Printed in the United States of America

ISBN 978-1-62534-184-6 (paper); 183-9 (hardcover)

Designed by Sally Nichols
Set in Adobe Minion Pro
Printed and bound by Sheridan Books, Inc.


Library of Congress Cataloging-in-Publication Data

Macieski, Robert, 1957– author.
Picturing class : Lewis W. Hine photographs child labor in New England / Robert Macieski.
pages cm
Includes bibliographical references and index.
ISBN 978-1-62534-184-6 (pbk. : alk. paper) — ISBN 978-1-62534-183-9 (hardcover : alk. paper)
1. Hine, Lewis Wickes, 1874–1940. 2. Child labor—New England—History—20th century.
3. Child labor—New England—History—20th century—Pictorial works. 4. Children—New
England—Social conditions—20th century. 5. Documentary photography—United States. I.
Hine, Lewis Wickes, 1874–1940, photographer. II. Title.
HD6250.U4N3663 2015
331.3′1097409041—dc23
                                                2015027973


British Library Cataloguing-in-Publication Data
A catalogue record for this book is available from the British Library.


All of the Lewis Hine photographs reproduced are from the
Library of Congress, Prints & Photographs Division,
National Child Labor Committee Collection.
Information about the collection is available online at
http://loc.gov/pictures/collection/nclc/.

*For my sister Katy*

# Contents

Acknowledgments    ix

1. Lewis Hine in New England    1

2. Street Trades    21

3. Textiles    48

4. Exhibiting Child Labor    77

5. Sardines    103

6. Seasonal and Family Labor    123

7. Exhibiting Child Welfare    153

8. Homework    179

9. Working-Class Communities    206

10. Trades and Vocational Education    235

Notes    265

Index    289

# Acknowledgments

This book began with an invitation from John Frisbee, director of the New Hampshire Historical Society, to join with him on an exhibition of Lewis W. Hine's photography from northern New England. I immediately accepted, thrilled at the opportunity to work with Hine's magnificent photographs again. There are many wonderful people at the Historical Society but I am particularly grateful to have worked with D. B. Garvin and Wes Balla. D. B.'s local knowledge and good humor make for an irresistible combination, and Wes is the consummate professional. John died shortly after the exhibition concluded. I grieve the premature loss of a friend and mentor.

I first learned about Lewis Hine in graduate school from Judith E. Smith, who sent me to Rhode Island as a research assistant to look for images for her book *Family Connections A History of Italian and Jewish Immigrant Lives in Providence, Rhode Island, 1900–1940*. My first stop was Slater Mill Historic Site. The curator introduced me to Hine's Rhode Island photographs. I was hooked immediately.

As fortune would have it, a decade later I was the curator of that collection at Slater Mill, enthusiastically showing the Hine's work to interpret immigration and industrialization in the Blackstone River valley, as my predecessor, Ruth Macaulay, had done. Christopher Chadbourne & Associates, a design firm, created an immersion setting of homework, and the Rhode Island Historic Society used those images in an exhibition at the new Museum of Work and Culture in Woonsocket and re-created a scene from one of Hine's depictions of homework.

I must thank my friends and comrades in the Rhode Island Labor History Society, particularly Scott Molloy, doyen of Rhode Island labor history. He and the now late Chuck Schwartz directed me to many documents and artifacts of child labor. Together with Rick Brooks and Holly Bagley, we created the first Rhode Island Labor & Ethnic Heritage Festival, a magnificent, joyous event during which we exhibited these photographs to ten thousand visitors on a beautiful sunny Labor Day at the Slater Mill.

At the University of New Hampshire at Manchester, I want to thank the librarians

Carolyn Gamtso, Annie Donahue, Ginger Borase, Rachel Blair Vogt, and Cindy Tremblay for their unfailing good cheer no matter what request I brought to them. My appreciation also goes to my colleague Barbara Jago for reading a draft, offering comments, and cheering me along. When I asked Interim Dean Michael Hickey for support to aid in creating an index for this book, his reply was "Where there's a will, there's a way," and for that I am deeply grateful. Sue Walsh, Dan Reagan, and Kathy Braun found the way. Thank you all.

Many thanks to Hetty Startup, who gave counsel early on in this undertaking and helped me to think visually about labor history. My friend Jim O'Brien read an early draft and offered valuable insights into the organization of the book. For thirty years I have anticipated writing a book that Jim could index for me. The time has finally arrived.

I can't thank Ardis Cameron enough for reading and commenting on a draft of this book. Throughout graduate school Ardis was a friend and an inspiration. Her good humor and generosity of spirit continue to shape our friendship.

I first presented this work to the faculty of Moscow State University, where in 2004, I was awarded the Fulbright appointment of Nikolia V. Sivachev Distinguished Professor of American History. The hospitality of Eugene Yazkov and his wife Marina, a Stalingrad survivor, made this time memorable beyond imagining. Yuri N. Rogoulev generously fed my family and toured us around Moscow to show us Soviet playgrounds, athletic fields, and cultural centers for children. Constantine Beloruchev graciously helped in all stages of my stay in the city. Thank you for your friendship.

Throughout the writing of this book, my bandmates played a unique role of keeping me grounded and bringing the joy of music into my life. Thank you Rick Fortin, Fjodor Heer, Roger Dodger, and Rafal Dybizbanski.

I learned about child labor before I learned of Lewis Hine, from my grandmother, a child laborer. I asked her once why she, a cultural Jew, was so committed to her Christmas clubs, which she had for each grandchild and weekly deposited $3 into each account. She said she had never had a toy as a child, which explained a lot of things. My grandmother went on to tell me about going off to work at the age of twelve. She worked every day until she could work no more. She lived the life of the children I was studying. This gave me a particular affinity with Hine's subjects. I must add my gratitude for my grandmother's contribution to my education, an opportunity she was denied. She sent $5 a week throughout my college years to buy me a pastrami sandwich, nurturing my body as well as my mind.

My brother, Daniel, and sisters, Lisa, Joanne, Katy, and Jessica, made my own childhood a happy one, and their support has been unwavering, as my parents' has been. An autodidact, my father would have been immensely proud of this accomplishment.

I regret that he is not here to enjoy it. Mom, thanks for making good things happen. I love you all.

The irony was not lost on me that as I was writing about the horrors of child labor, I was pushing my sons, Sam and Ben, to go out and get themselves jobs. Growing up they had little in common with these youthful laborers. Sam and Ben made this a longer journey, but also a more interesting one. Thanks, boys.

Finally, I thank my partner, Glenda Coffin, whose sacrifices went unspoken and whose encouragement throughout this project was unceasing. In addition to creating magnificent artwork, Glenda's great gift has been to share with me her love.

# PICTURING CLASS

FIGURE 1.1. "Addie Card, 12 years. Spinner in North Pormal [Pownal] Cotton Mill." (Hine 1050)

## CHAPTER 1

# Lewis Hine in New England

—

IN THE SUMMER OF 1910, Lewis Wickes Hine traveled through the Green Mountain State on behalf of the National Child Labor Committee, searching for opportunities to photograph children at work. This was Hine's second trip to Vermont in as many years, though the state was hardly on anyone's list of worst offenders of child labor laws. Quite the contrary, most Americans perceived Vermont as a largely rural state, punctuated by modest pockets of industry perhaps, but defined more by farms, fields, and forests than by factories, mines, or mills.[1] That August, however, Hine sought out those scattered sites of industry, moving along the state's waterways through the rolling hills and protected hamlets, searching for images of the state's children at work. It was his first fieldwork since the previous December.

Stopping in textile communities such as North Pownal, Bennington, Proctor, Winooski, and Burlington, Hine "discovered" a young girl who became one of his most dramatic photographic subjects. Addie Laird, of North Pownal, was working as a spinner in the North Pownal Cotton Mill, situated on the banks of the meandering Hoosic River.[2] When Hine photographed Addie for the National Child Labor Committee, he created a compelling portrait that conveyed the cost and consequences of child labor. Hine positioned Addie slightly off center in front of a spinning frame (fig. 1.1). Her filthy bare feet are planted firmly on the floor, her smocklike dress spotted with grease, oil, and dirt stains. Her pocket bulges with things mill workers needed: rags, perhaps tools, scissors, and the like. In case there was any doubt as to her role in the factory, Hine posed Addie with her arm reaching back awkwardly to the spinning frame, connecting her physically, as well as metaphorically, to the machine we assume she worked daily. The innocence of her face and the frailty of her limbs contrast starkly with all that surrounds her. Her head is level with the spinning frame, intersecting the rows of roving and thread, whose lightness in the enveloping dark create horizontal elevations that frame and hold Addie in the composition.

To resist any misreading, Hine's captioning fixes his message to the image. Hine identifies Addie as a spinner in the North Pownal Cotton Mill and notes that the "girls in mill say she is ten years" old; however, Addie "admitted" to Hine that she was twelve. Hine also notes that Addie told him she started working during the school vacation but had subsequently decided to "stay." Hine suggests that mill work has already taken a toll on Addie's health, diagnosing her in the caption of another photograph as an "anaemic little spinner."[3] Addie appears as well in a group portrait of fifteen boys and girls, all between twelve and fifteen years old, gathered outside the mill. Hine managed to collect their names and ages to include in his caption, which announces that, despite unemployment and "slack" work for adults, all the children work inside. The image tells us that Addie was not alone—she was part of a community of young workers.

The portrait of Addie Laird is one of Lewis W. Hine's best-known and most enduring images. In the century since it was taken, Addie's photograph has been deployed widely, in documentary and fictional works, on websites, and in videos, films, museum exhibits, and commercial advertisements. Like Marilyn and Elvis, Addie was commemorated by the U.S. Postal Service, which in 1998 issued a thirty-two-cent stamp bearing her image. A joint act of the Vermont legislature that year resolved that Hine's photograph of Addie Laird stood for the "thousands of children [who] toiled horrendously long hours in the manufacturing plants and mills of New England." Furthermore, according to the act, Addie represented immigrant industrial New England, even if she herself was not foreign-born: "the rural southern Vermont village of North Pownal, Vermont, was home to a substantial French-Canadian immigrant population that was drawn by the presence of a large cotton mill that offered employment opportunities." The mill in North Pownal, the act noted, "as was typical of its regional counterparts, employed many youngsters, particularly girls, who worked for long hours, in dangerous conditions and for extremely low wages." On that day in 1910, "Lewis Hine photographed a young girl named Addie Laird who spun cotton at the North Pownal mill for merely pennies a day," creating a "haunting pictorial portrait of an American child doomed to the misery of the cotton mill . . . a tragic aspect of our state's proud industrial history."[4] The encounter was frozen in time; Hine had immortalized Addie.

In captioning the photograph, however, Hine misidentified his subject. Thanks to the perseverance of the author Elizabeth Winthrop (whose novel *Counting on Grace* is based on the photograph) and the diligence of her research associate, Joe Manning, we now know that Addie's last name was Card, rather than Laird. Her life script,

Winthrop and Manning found, did not deviate significantly from what one might imagine for her based on Hine's images and captions. As is true of the majority of the poor, Addie's misfortunes began at birth. We know that she was born into poverty in Pownal, Vermont, in 1897, but we know nothing of her father and only that her mother died of peritonitis when Addie was two years old. At eight, she entered the mills for the first time.

The 1910 census helps fill in some information about Addie. At the time of the census, Addie Card was twelve years old and living in the home of her sixty-two-year-old grandmother, Adelaide Harris, along with Harris's twenty-six-year-old son, Trevor, and five other orphaned or abandoned grandchildren, including Addie's sister Anna Card, age fourteen, and Thomas, Frank, Susie, and Malden Harris, ages eleven, nine, four, and two, respectively. The only employed people in the household were Adelaide's son, a teamster at the cotton mill, and Addie and Anna, both spinners at the cotton mill. Through Winthrop's efforts, we have learned that at the age of seventeen Addie married Edward Hatch, also a spinner, and the two had a daughter. In their divorce in 1925, Addie lost custody of her child. Shortly afterward Addie was married a second time, to Ernest Lavigne, and together they adopted a girl, the "illegitimate" daughter of a Portuguese sailor. This second marriage was never a particularly happy one either. Ernest was an alcoholic and left Addie a widow in the late 1960s. When Addie died at ninety-four, she was collecting a Social Security check and living in low-income housing.[5] She never knew that the photograph taken by Lewis Hine had immortalized her as a quintessential emblem of child labor in the United States.

Addie Card served as an icon of the child labor movement. Hine's photograph of her was part of a new type of representation of the region—images of young workers toiling, of industry churning, of immigrants making their way in the world. These photographs were not of quaint New England farm life, domestic scenes, or rustic seaports but of the newly formed proletariat, and they inspired the professional, middle-class Progressives of the day to work for reform in various arenas—housing, public health, education—in seeking to rescue the likes of Addie from an industrial future. Addie's photograph joined hundreds of pictures in what became a massive archive of child labor images that now resides in the Library of Congress. The archive, and the values that supported it, magnified the power of images like the one of Addie.

Lewis Hine kept producing images like Addie's in the hope that they would prick the conscience of the nation and bring about the eradication of child labor. His photographs certainly weighed heavily on the minds and hearts of Progressive reformers, who were aggressively engaged in activity that promised to elevate the lives of children. They saw Hine's photographs as substantiation of the necessity for reform and used them as empirical evidence in a public campaign to strengthen the regulations against child labor. Contrary, however, to the historian George Dimock's argument that Hine's photographic subjects are anonymous, merely case studies or work types, I believe that the image of Addie presents her as a "real" person—a specific and autonomous individual, the subject of her own portrait, an actor in her own life.[6]

Progressive reformers shaped the campaign against child labor; their main organization was the National Child Labor Committee (NCLC), formed in 1904. The NCLC's board of directors included middle-class reformers such as Florence Kelley, the first general secretary of the National Consumers League; the pioneer settlement house worker Jane Addams; and Felix Adler, the founder of the Ethical Culture movement, among others. They were joined in the larger movement by trade unionists, socialists, ministers of the social gospel, and, most important, club women. For ten years beginning in 1908, Lewis W. Hine worked for the National Child Labor Committee (NCLC), traveling around the country to document children's working conditions. In New England, the documentation of child labor brought forth a multitude of reformers who sought to bring about incremental change. In the southern states the reform question remained unanswered, as many business leaders feared the "entering wedge" of labor legislation and resisted any effort to regulate child labor. Members of New England's new professional middle class, however, stood ready to put their ideas about social work, housing, planning, and recreation to work. Progressive professionals shaped reform throughout the region. Hine fit in to their high-minded mission with ease.

## PROGRESSIVE EDUCATION

Lewis Wicks Hine (1874–1940) was himself a product of Progressive education. He was born in Oshkosh, Wisconsin, to Douglas Hull and Sarah Hayes Hine, both of New England stock; Hine's father operated a coffee shop and restaurant in Oshkosh until his untimely death left Lewis to fend for his mother and himself. Hine's life took a most fortunate turn when he decided to resume his education and began taking university extension classes. He soon came to the attention of Frank A. Manny, head of the Department of Psychology and Education at the State Normal School in Oshkosh. Manny saw something in Hine and encouraged him to study education at the University of Chicago. Chicago was at the time a vibrant laboratory of Progressive reform, and the University of Chicago a fertile intellectual incubator for the Progressive movement. At the university Hine embarked upon his life's path. Under the luminous influence, if not direct tutelage, of John Dewey, the philosopher and educational reformer, Hine began his training in education. When Manny was appointed principal of the Ethical Culture School in New York City, he took Hine and other protégés along and engaged Hine as an assistant teacher of nature studies and geography. Though Hine would later return to the University of Chicago for a summer session, he enrolled in the School of Education at New York University (NYU) and continued his studies. In

1905 he received his masters in pedagogy from NYU; the following year he enrolled in Columbia University's Graduate School of Arts and Sciences to study sociology.[7]

At the Ethical Culture School, Hine formed a camera club and taught a camera class; both in class and in the club, Hine joined students in sharing "photographic experiences" that nurtured a "sharpened vision." Hine taught students that "real photography" is more than a matter of a "lucky hit"—it involves "intelligent, patient effort" in selecting a subject and structuring a composition. Hine praised the utilitarian benefits of photography, which provided potential work or "wholesome hobbies" for the "city child," but he argued that the true educational value of photography was in fostering a way of engaging and understanding the world. Hine believed that the impact of photography was dialectical: it could assist in teaching us to see and in shaping what we know, both of which are important. As the art critic, novelist, and poet John Berger has argued, "what we know and what we believe" affect "the way we see things." In the "last analysis," Hine contended, "good photography is a question of art" and the development of an "artist's point of view," from which the "social effects are manifold."[8] From the earliest, Hine linked his photographs with their potential social impact.

Hine first put this philosophy of education and photography to work at New York Harbor's Ellis Island, the nation's largest and busiest immigrant-processing center in the first half of the twentieth century. Beginning perhaps as early as 1903, he journeyed to Ellis Island and, with his 5 x 7 camera mounted on a shaky tripod, recorded the social phenomenon of immigration, which was much discussed, little understood, and widely disdained. The camera, Hine believed, could provide a fresh perspective on the process. His photographs helped to turn a "faceless horde" into nameless but undoubtedly human individuals and families. Hine's technical sophistication was minimal: using a rapid rectilinear lens and an old-fashioned shutter with plunger and magnesium flash powder to light the dim halls of the massive Ellis Island structures, Hine produced approximately two hundred images that are some of the most powerful and important pictures in the history of American photography. What *was* sophisticated was the photographic perspective that Hine began to develop there and that he refined over the next four decades. Hine's friend Walter Rosenblum believed that Ellis Island was the "crucible that formed Hine, gave him direction, and schooled him for what was to follow."[9]

## THE PITTSBURGH SURVEY

The Pittsburgh Survey project began Hine's transition into being a professional photographer, one who embraced the new approaches of the emerging field of social work. Hine's sympathetic photographs of recent arrivals to Ellis Island gave a humanizing dimension to a pressing social concern, suggesting the power of photography to reveal a fresh perspective. Hine's work on the Pittsburgh Survey in 1907–8 affirmed this idea and further taught him that photographs could be persuasive instruments for social change. Sponsored by the Russell Sage Foundation, the Pittsburgh Survey was an unprecedented and detailed study of workers and the city in which they lived. It was designed "to get at the facts of social conditions and to put those facts before the public in ways that will count." Years after the project, Paul Kellogg, the editor of the social work journal the *Survey,* described the new methods of putting this information before the public, writing in a letter that when he "launched the Pittsburgh Survey back in 1907–8, we broke with the old stereotypes of social and economic investigations. We reinforced our text with things that spoke to the eye—drawings in charts, graphs, maps and designs from engineers—and employed not only photographs but pastels and even sculpture, using Rodin's *Puddler,* for example, to visualize the meager payments made for the loss of an eye, an arm, a leg, in the industrial accidents of the steel mills. . . . And we turned to Lewis W. Hine." Hine, he concluded, "is the pioneer in social photography of this sort."[10]

The Pittsburgh Survey served as a model "scientific social survey" for Progressive reformers and offered Hine entrance into an institutional world and ideological context ideally suited to his work. Hine joined networks of professionals who worked within dozens of professional reform associations, placing their faith in reason and systematic methods of investigation; these individuals and organizations shared their findings in popular media such as newspapers, magazines, books, pamphlets, and other publications and through exhibits, conferences, and lantern shows. The "authenticity" of the photograph also fit well with the Progressives' emphasis on gathering theoretically neutral and nonpartisan data that could help identify the root causes of a social problem, if not also provide solutions for its resolution—evidence that could bring order to the seeming chaos of modern life.[11]

Hine's aesthetic sensibilities began to become clear and identifiable in his Pittsburgh Survey work, in which his treatment of his subjects demonstrates recurrent themes. His Pittsburgh photographs, as with the Ellis Island images, portray the immigrant workers in sympathetic terms. His subjects were usually posed facing the camera, with their heads and shoulders framing the image, and they often appear to be looking directly at the viewer, making intimate contact despite the interruption of time and space. The impact called for engagement between viewer and subject. No matter how disoriented his subjects were personally or how depressed their social conditions, Hine was able to capture in them a sympathetic human quality, a life-affirming dignity that no hardship could completely destroy. Eyes meet eyes, forging an insoluble bond between human beings.

The Pittsburgh Survey provided a magnificent opportunity for Hine to establish his work before the public and an equally valuable chance to develop important contacts among Progressives. The shift in the reform movement from an emphasis on charity and flawed individual characters to a more systemic investigation of social conditions and a rational approach embraced by professional social work suited Hine well. The social survey assumed preeminence among social workers. Though Hine's photographs often focused on individuals or small groups, through their repetition his images provided powerful evidence of social conditions. Through the *Survey* magazine, Hine also established important professional contacts. Hine developed a sound working relationship with Paul Kellogg, editor of the *Survey,* and Arthur Kellogg, Paul's brother and the managing editor of the magazine, which supported Hine through much of his life. In these years, Hine also worked with the child labor reformer Florence Kelley; with Lillian Wald, a nurse, founder of the Henry Street Settlement, and cofounder of the National Association for the Advancement of Colored People; with John Spargo, a public intellectual affiliated with the Socialist Party of America; and with others who came to appreciate the value of Hine's work in the cause of social welfare and reform. The Pittsburgh Survey was the most important example of modern practices in the emerging field of social work, and Hine's participation effectively positioned him within the orbit of social reform.

## SOCIAL PHOTOGRAPHY

In 1909, Hine addressed the National Conference of Charities and Correction on the subject of "Social Photography; How the Camera May Help in the Social Uplift." This was Hine's only public statement on social photography and the use of photographic images to promote social reform.[12] In June of the previous year, Hine had first advertised his services as a social photographer in *Charities and the Commons,* the precursor to the *Survey,* but his ad said nothing about the meaning of social photography or the function of the social photographer. At the Charities Conference, however, Hine made his case for the central role of photography in social reform movements. To be successful, Hine argued, the campaign against child labor, like other reform efforts, must arouse the "public sympathy." Sympathy, he observed, was the fulcrum of change. Sympathy need not be left to chance; rather, it was up to reformers as the "Servants of the Common Good to educate and direct public opinion." And such servants, Hine confessed, were "only beginning to realize the innumerable methods of reaching this great public." Foremost among these methods was the social photograph, later known as the documentary photograph. As the historian Maren Stange has persuasively argued, Hine earned his reputation as a skilled practitioner of social photography, the new documentary form, through his work on the Pittsburgh Survey.[13]

Hine hailed modern techniques of advertising as vital tools for reaching the "great public" with his sympathetic portraits. Quoting from a recent article in *Collier's* magazine, Hine claimed, "Advertising is art; it is literature; it is invention," but he also acknowledged the difficulty of being heard among the cacophony of voices when "all are tooting the loud bazoo." Unlike enterprising manufacturers selling their new wares, the "social worker, with the most human, living material as his stock in trade," must address the problems and activities of life, with all "their possibilities of human appeal." Faced with such challenges, it was imperative, therefore, that the social worker "eagerly grasp at such opportunities to play upon the public sympathies of his customers as are afforded by the camera."[14] The social photographer needed to use the camera to educate the public sufficiently to the reformers' viewpoint to evoke the desired sympathy.

Graphic representations, Hine maintained, are capable of creating "a symbol that brings one immediately into close touch with reality." He believed that every image "speaks a language learned early in the race and in the individual," becoming "the language of all nationalities and all ages." The tremendous growth in the use of illustrations in newspapers, books, and exhibits in the first decade of the twentieth century was precisely because they could transcend linguistic limitations and confer additional authority to the author or the cause. Hine argued that pictures "tell a story packed into the most condensed and vital form" and derive their power in ways that can be more influential than the reality would be; he was confident that photographs could improve on reality itself because "in pictures, the non-essential and conflicting interests have been eliminated." Photographs told a story by transforming images into graphic symbols and, through the storyteller's framing ability, evoking the desired sentiment.

In his effort to explain to his audience the potential power of graphic representations, Hine recognized and advocated the enhanced influence gained by linking words with images. Hine understood that photographs have power on their own terms, but he also acknowledged the ways in which creative captioning can reinforce an image and expand the photographer's reach for public sympathy. For example, discussing one of his photographs of a tiny spinner in a Carolina cotton mill, Hine told his audience that it "makes an appeal." But if the image were joined with "one of those social pen-pictures" written by the novelist Victor Hugo, the meaning of the image, and its ability to arouse sympathy would be amplified. With photographs "thus sympathetically interpreted, what a lever we have for the social uplift."[15] Meaning in Hine's photographs is the product of synchronizing image and text. Hine captioned the vast majority of his photographs and included many of his captions in his picture essays, photo stories, and "Time Exposures," efforts to tell narratives through a combination of words and images.

If a photograph could bring individuals in close touch with reality, Hine was careful to differentiate the image from reality. He acknowledged to his audience that photographs, perhaps more than any other graphic form, possess a "realism" of their own, leading to a false but powerful assumption among many in the public "that the photograph cannot falsify." In one of his most frequently quoted statements, Hine cautioned against "unbounded faith in the integrity of the photograph," because "while photographs may not lie, liars may photograph."[16] In using the camera as an instrument for the "revelation of truth," the photographer had to guard against "bad habits" and "yellow-photography." Photographing for a cause posed particular temptations that threatened the veracity of the image.

The integrity of the photograph was essential to its value as evidence. Certainly Hine understood that images can be manipulated and that not all photographs are as they appear. Hine explained that a leader in social work had told him that "photographs had been faked so much they were of no use to the work." *Survey* editor Paul Kellogg reassured Hine that was not the case with his recent photographs of child labor; they "would stand as evidence in any court of law." It is not clear, however, whether at that point in his career Hine understood how difficult it might be for a photographer advocating a cause to keep the "bad habits" at bay and maintain the close touch with reality he spoke of to his audience of social workers. In the photographic assignment Hine had most recently completed, photographing children who were working the street trades in Connecticut, he had exhibited his emerging skill at editing out the "non-essential and conflicting interests" in his images to achieve clarity of meaning. The meaning Hine sought was empathy. In his images and in his captioning, Hine tried to balance the sometimes conflicting demands of evoking sympathy and representing reality.

### AMERICAN CHILDHOOD

Hine was photographing at a time when ideas about childhood were changing. The topic of childhood is one of perennial concern to Americans, and it is heavily invested with well-meaning prejudice and commonly held misunderstandings. The historian Steven Mintz, in his survey of the topic, *Huck's Raft,* argued that the history of childhood is obscured by a "series of myths" about a happy, "carefree childhood" and a loving, welcoming "home." Many misunderstandings stem from the belief that childhood is basically a "biological phenomenon," similar for all—"a status transcending class, ethnicity, and gender"—rather than a "life stage" shaped by particular social and cultural circumstances rooted in specific times and places. The literature was and still is polarized between the notion that childhood is on a path of perpetual progress and improvement and, conversely, the belief that American childhood is fast disappearing, suffering an inauspicious decline brought on by those who force children to grow up too quickly.[17]

Ideas about childhood were changing dramatically in the early part of the twentieth century. Edward Devine, secretary of the Charity Organization Society, spelled out the era's new thinking. First and foremost, every child had a right to be born, whether or not the child was wanted. The right to be born was followed by the right, as he put it, to grow up. A third element in the new view of childhood was that the child had a right to be happy, "even in school." The nineteenth-century educator Friedrich Froebel staked out the claim that children had the right to be joyful in learning. At one of the meetings of the NCLC, Jane Addams expressed her belief that future generations would be embarrassed by the treatment of laboring children. Children had the new right to "the glorious fulfillment of enjoyment for which children are by nature adapted, and by their creator intended." Protecting children from work was absolutely essential to their having a happy childhood. The new view maintained that the child had a right to become a useful member of society, and this meant expanding vocational education. Devine called for making the school day longer and the curriculum more varied, with additional vocational opportunities. Finally, Devine articulated the child's right to do better than his or her parents: children had the right "not only to be protected against degeneracy" but also "to progress." The fundamental elements of what Devine called "the right view of the child were: "normal birth, physical protection, joyous infancy, useful education and an ever fuller inheritance of the accumulated riches of civilization."[18] Hine embraced these ideals, and they served to shape his images.

These mythologies of American childhood were the filters through which Hine's images are most often seen and against which they obtain their power and influence. As the historian James Guimond argues in *American Photography and the American Dream*, the story represented by Hine's images serves as a counter-narrative to the "American dream."[19] Hine's photographs document the betrayal of New England values that placed a premium on education, telling a story of harsh industrial conditions and conveying the price of economic development and, perhaps, the callousness of a society that allowed its young to labor so early in life. In this way, Hine's photographs act as a counter to the mythologies of childhood purity, innocence, and joy that were dominant in the culture of the era. Viewed through these cultural filters, the images tell a story of transgression, a sign of the failure to parent and protect the young.

Meaning is derived from interpretation, which itself is influenced both by the social and cultural context of a photograph's creation and by the cultural context of the photograph's observation or viewing. Invariably, there are those who invest photographs with a more sedentary power to mirror society, to reproduce on paper an immutable truth existing in society. Photographic meaning is rarely so singular or static. The American studies scholar Miles Orvell noted the "ambiguous quality" of photographs, which are "both objective and subjective at the same time, a window into 'reality' and a constructed language." Orvell echoed the theorist Roland Barthes's semiotic assessment, which claimed that photographs are coded so that "how the subject is shown, the various cultural and aesthetic codes . . . tell us more fully what it means."[20] Hine's photography navigated those cultural and aesthetic codes to generate meaning. While he did not invent the codes, he did play to them and reinforce them.

Although good fortune clearly played a role in their production, Hine's photographs were not an accident—he constructed the portraits, working with the apparent cooperation of his subjects. These images reflect the "contact zone" between photographer and subject.[21] This is the place where Hine and his subjects—the reformer and those to be reformed—were joined, despite their divergent origins and status. Power relations were clearly uneven in these meetings, with Hine's age, gender, education, and social class giving him the advantage, but however imbalanced their relationship with Hine, the children were not merely passive and easily manipulated: they played an active role in shaping the images.

## SOCIOLOGY AND A CAMERA

With his education and training, Hine wasn't simply a photographer. He was a sociologist with a camera. Hine (fig. 1.2) was concerned primarily with portraying social reality, but he didn't just snap the shutter, capturing random cross sections of his subjects' lives. In framing his images, in selecting his subjects and their surroundings, Hine was *constructing* social reality. His images became part of the contemporary discussion of childhood, particularly the issue of child labor. In each image Hine framed the values and ideals that guided the child labor movement. By looking at Hine's photographs as social constructions expressing reformers' values, we can examine them in light of Progressive discourse on class, gender, and race. The American studies professor Laura Wexler has argued that "the comparative dearth of critical attention to the social productions of the photographic image is a class- and race-based form of cultural domination."[22] The production of Hine's photographs was complicated by the addition of text to his images, but the text also revealed the intent of his efforts.

If Hine had an ideology, it could best be described as that of a social worker. Kate Sampsell-Willmann maintains in *Lewis Hine as Social Critic* that Hine turned to pragmatism, which informed his image making, but if he did so, it was later in his life, after he had worked against child labor. She is much more on target when she argues that Hine became a "professional photographer within the Progressive world of social work and social causes."[23] In the introduction to *The Spirit of Social Work,* Edward Devine described the social worker as one who

> in any relation of life, professional, industrial, political, educational, or domestic; whether on salary or as a volunteer; whether on his own individual account or as

FIGURE 1.2. Portrait of Lewis Hine, ca. 1930. George Eastman House, gift of the Photo League, New York: excollection Lewis Wickes Hine 78:1059:0046.

a part of an organized movement, is working consciously, according to his light intelligently, and according to his strength persistently, for the promotion of the common welfare—the common welfare as distinct from that of a party or a class or a sect or a business interest or a particular institution or a family or an individual. . . . [He] takes the social point of view and brings himself consciously or unconsciously into the ranks of the nation's social workers.

The discipline of social work was broadly conceived. In the first decade of the century, Devine reported, "the multitude of social workers, engaged in various occupations, enrolled under various banners, have made mutual discovery of one another's existence." Having become aware of one another's shared aims and aspirations, this multitude discovered that their "mutual interest in social work . . . differentiates them not only from the exploiter but from the neutral and indifferent member of society."[24] Hine was among that multitude. Social work was a commanding ethos in Hine's life as he photographed for the NCLC.

### CLASS AND IDEAS OF REFORM

Hine's photograph of Addie Card remains a powerful reminder of the excesses of industrial capitalism in the early twentieth century. The iconic image is recognized internationally, symbolizing those excesses and the cheapness of life in the world governed by profit. The image was a visual record representing a moment in time, but it has assumed what the French literary theorist and semiotician Roland Barthes called "the reality effect," enabling it to take on a life far larger and more lasting than the conditions it presents.[25] Hine's photographs ensured that Addie and her peers would exist beyond their own time. Because of these images, children who labored arduously day in and day out in U.S. factories, workshops, and fields exist in historical memory, if not in the public accounting of the costs of capitalism and the role children played, and continue to play, in the accumulation of capital.

Children should figure more centrally in the history of industrialization in the United States. Child labor was not an aberration of industrial capitalism but one of its core products.[26] Child labor accompanied mechanization. As Karl Marx famously wrote of the division of labor generated by capitalist methods of production, "In so far as machinery dispenses with muscular power, it becomes a means of employing laborers of slight muscular strength, and those whose bodily development is incomplete, but whose limbs are all the more supple. The labor of women and children was, therefore, the first thing sought for by capitalists who used machinery."[27] Alexander Hamilton made a similar statement in his *Report on Manufactures* (1791).[28] The employment of children was widely practiced in the United States, as it was in other capitalist countries, most notably in Britain. The 1900 U.S. census recorded that close to 2 million children labored for wages outside the home.[29] Children found their way from the countryside (or from abroad) into the nation's cities and factories. Hine's photograph of Addie is an expression of immiseration. But Hine did more than capture the individual fate of one child; he created an archive of images of exploitation and envisioned that the cumulative weight of photographic evidence he was gathering would turn public sentiment toward the further regulation of child labor.

Sampsell-Willmann asserts that Hine's relationship to work was "nuanced" and that "he became a member of the working classes—a skilled artisan."[30] To the contrary, I treat Hine as a middle-class reformer. Hine was a salaried worker, under the authority of higher-ups to be sure, but sufficiently independent to exercise his own judgment and carry his own byline. He allied himself with the Progressive forces and pursued the middle-class agenda of child labor reform, advocating amelioration of conditions for America's youngest workers. His job while working for the National Child Labor Committee—one that he became very good at—was to contrast the destructive effect of child labor on the coming generation with a valorized middle-class notion of childhood that embraced new ideas about adolescence put forth by writers such as G. Stanley Hall, the first president of the American Psychological Association, the first president of Clark University, and a leader in the child study movement. (Hall was instrumental in the development of educational psychology and attempted to determine how adolescents might best be educated.[31]) Hine did not offer a systemic critique of capitalism; he merely sought a more humane form. He was empathic with his subjects: he genuinely sought their uplift, but he was not in mystic accord with them.[32] Hine was too professional, too objective, to take a position in solidarity with them. He would fight on their behalf, but he was a reformer, not a radical.

Hine's vast corpus was very much a product of the fervent reform activity of the Progressive movement. Hine joined the child labor movement just as it was coalescing. He painted his portrait of American childhood image by image, optimistically believing that his work's cumulative weight would serve as a lever for change, acting as a catalyst for legislative reform of unjust social conditions. He viewed each image as having the potential to raise consciousness about social conditions and expose the vast gulf between the rhetoric and reality of American life. Hine told one audience, "For many years I have followed the procession of child workers winding through a thousand industrial communities from the canneries of Maine to the fields of Texas. I have heard their tragic stories, watched their cramped lives, and seen their fruitless struggles in the industrial game where the odds are all against them. I wish I could give you a bird's-eye view of my varied experience."[33] Through his photographs, Hine intended to do just that.

## CHAPTERS AND COMPANION PORTRAITS

The individual critique of society present in each photograph can also foster a collective critique if the images are viewed in sequence or as a group. Alan Trachtenberg, the leading scholar on Lewis Hine, has noted that while every image contained its own critical commentary based on its individual story, each "took its ultimate meaning from the larger story." According to Trachtenberg, writing in 1977, analysis of Hine's images requires a "most strenuous act of imagination on the part of Hine's audiences today—to see his individual images in the context of their companions."[34] Scholars have generally looked at Hine's photographs as single entities, as one might evaluate a painting or sculpture. Few have situated that individual image within a context of companion works. Fewer still have explored Hine's engagement in projects, the contexts Hine worked in, and the people he worked with.[35]

This book journeys with Lewis Hine through New England towns and cities as he photographed children at work. In each place he stopped he left a record—cumulatively, the images form a portrait of a generation of immigrant and working-class youth. Each chapter explores a different project situated in distinct places: newsies (newsboys) in New Haven, Hartford, and Bridgeport, Connecticut, in chapter 2; textile operatives in Rhode Island, Maine, Vermont, and New Hampshire in chapter 3; the Boston 1915 movement in chapter 4; sardine workers in Eastport, Maine, in chapter 5; cranberry and tobacco pickers on Cape Cod and in Connecticut in chapter 6; a Rhode Island child welfare exhibit in chapter 7; industrial homework in Boston and Providence in chapter 8; working-class communities in Lowell, Lawrence, Fall River, and New Bedford, Massachusetts, in chapter 9; and vocational training in Boston in chapter 10. The chapters follow Hine chronologically as he focused on different types of work, so that we can see how he matured in his craft and shifted emphasis, creating pictures of working-class children as seen through the camera lens of a middle-class reformer.[36]

## NEW ENGLAND AND CHILD LABOR

Hine photographed child labor across the country, but his work in New England not only violated local sensibilities and self-perceptions but did so in a region where Progressive efforts to uplift and eradicate social misery were active and widespread. Just think of it, the *Brooklyn Eagle* editorialized: "New England that freed the slave is taking work from its own people and sending it to the South because it can have it done there for almost nothing by infants." All New England states had histories of reforming child labor dating to the 1840s, and they expected more of themselves.[37] It was one thing for child labor to be rife in the South; it was another to see it widespread in New England. As Florence Kelley, the general secretary of the National Consumers League and a leading anti–child labor advocate, told a group of club women in 1902: "The people of the North could not well afford to throw stones at the South, because conditions here are bad enough." While many southern states didn't even have compulsory education laws, New England states had long prided themselves on the provision of education. But they could not rest on their laurels. Regarding the education of children, according to Kelley, the "six great Northern manufacturing states are gradually falling lower and lower in the scale and were in this respect worse off than the five great manufacturing states of the South—the two Carolinas, Georgia, Alabama, and Louisiana."[38] Because New England was more developed economically, more ethnically diverse, and more conducive to reform, Hine's work there is important.

Hine's work in the region was extensive and intensive and left little doubt about the pervasiveness of child labor. He photographed in most of the region's cities— Bridgeport, New Haven, and Hartford, Connecticut; Pawtucket, Providence, Woonsocket, and Central Falls, Rhode Island; Boston, Cambridge, Chicopee, Salem, Springfield, Somerville, and Worcester, Massachusetts; Lewiston and Sanford, Maine; Manchester, Dover, and Portsmouth, New Hampshire—and documented some of them in great depth, such as Fall River, New Bedford, Lawrence, and Lowell, Massachusetts. In nearly all of those cities, the foreign-born and the children of the foreign-born made up between two-thirds and three-quarters of the population.[39] Child labor defined a counter-narrative to the American dream, as the historian James Guimond has argued. While many might have agreed with the Puritan adage that "idle hands do the devil's work," few wanted the region to be known for child labor. The child labor that Hine captured on film exposed the region to ridicule and charges of hypocrisy, as New Englanders never hesitated to speak out against child labor elsewhere.[40] By 1910, the tables had turned somewhat regarding the eight-hour day, and it was the South asking, "How long is New England going to hold the rest of us back?"[41] Hine's photographs defined an emerging New England characterized by immigrants and urban life.

Hine expanded the vistas of visual culture in the region, introducing the industrial world though his photographs. Hine's New England images, such as his photograph of Addie, now stood alongside the more genteel work of John Singer Sargent, the "leading portrait painter of his generation," and *Girl in White* (1901) by the American impressionist William Merritt Chase, and the class differences were unmistakable. Among photographers, the pictorialists' portraiture was almost equally removed from the reality of working people. The fine art photographer Alfred Stieglitz included Gertrude Käsebier's photograph *Blessed Art Thou among Women* in *Camera Notes* (July 1900) and in the first issue of *Camera Works* (January 1903) because he considered it a work of art. But in that image, as in Käsebier's *Mother and Child* (1900), the child being introduced to the world is protected by her mother, not standing alone as Addie did. Class differences are starkly reflected in the presence or absence of a mother's supervision.[42]

Hine introduced modernity into depictions of New England, instilling his images with fragments of industrial capitalism and loading them with modern expectations for his photographic subjects. In this way his work departed from the common aesthetic expressions of the region. While Hine was capturing images of modern urban life, the art colonies of Old Lyme, Connecticut; Cornish, New Hampshire; Isle of Shoals, Ogunquit, and Kittery, Maine; Gloucester and Cape Cod in Massachusetts; and Newport, Rhode Island, were generating idyllic rural land- and seascapes, celebrating and striving for a colonial past. The painter Thomas Cole, regarded as the founder of the Hudson River school, had transformed the New England countryside, particularly the White Mountains, into scenic subjects for artists. The Boston Arts and Crafts movement reacted to modern industrialization by emphasizing traditional craftsmanship and frequent use of simple medieval, romantic, or folk styles of decoration. Among photographers, Wallace Nutting aggressively marketed the colonial past and helped to shape the Colonial Revival worldview. His books *New Hampshire Beautiful, Maine Beautiful, Massachusetts Beautiful,* and *New England Architecture* sold robustly. William Sumner Appleton joined in this embrace of the colonial past when he established the Society for the Preservation of New England Antiquities in 1910. The tourism industry contributed to this backward-looking perspective as well. As the historians Stephen Nissenbaum and Dona Brown have argued, "Promoters of tourism fashioned a self-consciously antiquated New England as an antidote to modernization."[43] Tourists traveled to rural destinations to escape the very type of urban scenes that Hine portrayed. By embracing the rural countryside as subject, the art of the region reflected the largely negative reaction to urban industrial life. Hine's photographs could only add to the unease that had led so many to seek refuge in an imagined colonial past to begin with. The children in Hine's images were not present in the region's artistic, literary, or cultural settings until Hine put them there.[44]

Hine's photographs of the immigrant working class questioned the popular iconography of New England. *The Puritan,* an iconic sculpture by Augustus Saint-Gaudens (an American sculptor of the Beaux-Arts generation) that celebrated Anglo-Saxon culture, could now be viewed in juxtaposition with Hine's image of Addie, the working-class textile operative—both of them symbols of New England. After Hine, bucolic depictions of the village green nestled around the church, with its white spire reaching heavenward, had to contend with portraits of mill villages in which children from many different countries trudged into the local factories to labor their lives away.

Hine documented the emerging ethnic diversity of New England and the rapid and unwanted influx of massive numbers of the foreign-born and their children. Hine's images were transgressive, contradicting the wholesome image of a quaint rural or seafaring New England past and violating middle-class ideals of childhood.

As the sons and daughters of Puritans met the waves of new immigrants from eastern and southern Europe, their reactions varied, from fear and loathing to acceptance and uplift. For some, such as the Portsmouth-born poet and novelist Thomas Bailey Aldrich, in his poem "The Unguarded Gates," the immigrants who came pouring into the country spoke in "accents of menace alien to our air" and threatened the nationalistic "white Goddess," Lady Liberty. Aldrich was among the many who supported the Immigration Restriction League, which urged Congress to "close the gates" and protect the "White republic" from Old World and Canadian immigrants. Others abandoned the cities and moved to the suburbs, where they would be protected from the newcomers and their foreign ways. Still others determined that if it was impossible to assimilate all of the foreign-born, an attempt should be made at least to socialize their children. This effort was carried out in schools and through the work of settlement houses and visiting nurses associations, boys' and girls' clubs, the YMCA and the YWCA, and an assortment of local organizations. Hine worked with these groups.

## NEW ENGLAND ARCHIVE

Lewis W. Hine photographed child labor from coast to coast and from North to South. He produced more than 5,000 images for the National Child Labor Committee, with over 1,500 created in New England. Among all the states he worked in, he photographed the most in Massachusetts by far, producing more than 1,010 pictures there during his tenure with the NCLC. Elsewhere in New England Hine took fewer photographs, but still in significant numbers: 167 in Connecticut; 108 in Rhode Island; 94 in Vermont; 72 in Maine; and 66 in New Hampshire. Hine gathered the names and addresses of dozens of the children he photographed and compiled hundreds of their purported comments, to which he added his own observations. Hine immortalized a generation of children by creating an archive unprecedented in the history of childhood and open to investigation by various disciplines. Our modern imaginings of child labor in the early part of the twentieth century are almost inseparable from the visual palette of child labor drawn from Hine's photographs. Scholars might debate Hine's skills or sophistication as a photographer, but it is undeniable that he created iconic images of children working.[45] And these images made it impossible to see child labor as an aberration, unique to one region, one industry, or one ethnic group.

Hine produced singular, iconic images, but it is the amplitude of his work, culminating in the creation of a vast archive, that is most impressive and least studied. Hine was picturing class in his work for the NCLC. He was photographing conditions in the capitalist economy, creating thousands of images that describe the labor conditions of the most vulnerable workers in the system: children. The resulting archive was intended to provide an authoritative body of evidence that would prove the existence and the extent of child labor. Not only is it an archive of class; it is also an archive of empathy, designed to evoke compassion with its thousands of depictions of children at work. The archive—containing more than five thousand photographs, captions, field reports, magazines, journals, exhibits, and ephemera—empowered the NCLC by overwhelming doubters and subduing critics, thus monopolizing the discourse on the subject.[46]

The archive of child labor portraits also subverted traditional cultural hierarchy. The overwhelming majority of Hine's photographs are portraits, frontal depictions of children alone and in groups. The visual theorist Allan Sekula has astutely observed that "photography subverted the privileges inherent in portraiture." Portraiture, once the preserve of the well-to-do, was now democratized by photography, so that Hine produced portraits that were upstarts in the cultural pecking order.[47] Certainly Hine's photographic portraiture was different from the established norms of painting portraiture. It created an alternative cultural hierarchy, inverted the social subjects, and disrupted the domestic contextualizing that framed traditional portraits of the elite. In New England, Hine introduced a new portrait, one of the young proletariat that was at odds with the latest middle-class views of childhood.

As effective as Hine's NCLC work was in prompting child labor reform one hundred years ago, and as compelling as it is today, one of the most interesting aspects of Hine's archive is that it contains only a small percentage of images in which the children, their parents, their adult coworkers, or their employers show any hesitancy, embarrassment, regret, or discomfort about who they are and what they are doing. They exhibit no unusual emotions or attitudes because working children were more the norm than an aberration. Usually Hine's subjects worked with him in the construction of the photographs. They followed directions well and sometimes seemed to know exactly what pose to strike. They cooperated in both small and large group pictures, as individuals and as families. They were not nearly as anonymous as scholars have contended. Hine was remarkably effective at gathering names and addresses even from children who spoke little English. Would they have been so cooperative if they had known they might be held up to public scrutiny?

Hine introduced new visions of New England into the region. Singularly or in sequence, Hine's photographs depicted some of the harshest elements of industrialism. The Colonial Revival style and paintings of pastoral landscapes had to make room for a new body of work, the archive of the National Child Labor Committee, which contained evidence that New England continued to produce child laborers despite a century of reform efforts. Though the region assumed a sectional superiority over the South, it was equally engaged in transgressing the cultural norms of the new middle class by retaining high levels of child labor. Let's follow Hine as he made his way through New England picturing class, describing in rich visual terms the nature of capitalist relations.

# CHAPTER 2

# Street Trades

―――

PROGRESSIVE PROFESSIONALS ENLIVENED NEW ENGLAND cities at the turn of the century with diverse efforts at municipal improvement. Rapid growth of these cities generated a cascade of urban concerns. Activists scrutinized building codes and promoted urban planning, surveyed crowded housing, inspected unhealthful living and hazardous working conditions, endorsed educational initiatives, and investigated commercial vice. Many citizens anxiously watched the growing concentrations of immigrants living in enclaves cut off from Americanizing influences. Given the multitude of uses people found for the streets, back yards, alleys, and dumps, Progressive activists were above all concerned about the safety of children turned out of their homes. Municipal professionals examined nearly all aspects of urban life and municipal government and condemned, among other things, the large number of children working in the streets.

During this time, middle-class women expanded their circles of friends and acquaintances to embrace professional connections and associations and stepped decisively into the public realm to work in behalf of others. In doing so, according to the historian Helene Silverberg, they "boldly entered political terrain once considered part of the male 'sphere.'" Through the General Federation of Women's Clubs, the National Women's Trade Union League, the National Consumers League, the National Child Labor Committee, and an array of local groups, middle-class women organized their efforts and promoted their causes. Lobbying state legislatures and testifying before legislative committees, these women laid the foundation of the welfare state by aiding the passage of legislation on women's work hours, child labor, and mothers' pensions.[1]

Lewis Hine worked intimately with these Progressive women, performing surveys and providing documentary evidence to validate the reformers' call for politicians and the state to attend to child welfare. He photographed child laborers through the lens of concern over city space and endeavored to present this evidence in the hope of transforming social conditions through the further regulation of child labor and reforming the urban landscape through the creation of alternative city spaces for children. Hine's photographs of the street trades generated apprehension about children being out at all hours, neglecting school and coming and going to places of questionable safety or morality. Middle-class reformers saw working-class urban spaces, and immigrant spaces in particular, as ominous and menacing: children in the city were surrounded by dangers. Hine constructed images that played to these social and cultural anxieties.

As a group, Progressive reformers embraced an environmental approach that invested crime, vice, and antisocial behavior in places as much as people. In the Progressive discourse, spaces possessed the values of those who inhabited them. Streets and alleys, back yards, trash dumps, and privies were imbued with moral and health dangers for unaccompanied children. Parks and playgrounds, on the other hand, particularly those that were supervised, served an uplifting function. Movie theaters and dance halls, many believed, seduced the young into immoral behavior, while boys' and girls' clubs, the YMCA and YWCA, and settlement houses offered alternative spaces that were safe and healthful. Streets were places of danger and foreboding, no more so than at night.[2] City planning designs, housing reform surveys, and vice commission reports conveyed a fear of promiscuous spaces and crowded and dirty environments, while at the same time expressing faith in the redemptive power of city spaces.

Middle-class Progressive reformers frequently analyzed city places in relation to the health of the body politic, using terms such as "congested," "infected," "diseased," and "virus" in their descriptions of urban conditions. This linguistic affinity for medical metaphors reflects the sharp distinctions in reformers' minds between licit and illicit use of city spaces. It also reflects the superior vantage point assumed by the new class of professionals who sought to remake urban life through city planning, housing reform, social welfare, public health, and recreational improvement. These professionals were society's social doctors, ready to operate as efficient, nonpartisan, scientific technicians and conduct the necessary surgery to sever the putrid, inflamed, unattended, and diseased parts from the urban patient.[3]

Hine's first work in New England involved photographing the street trades in the cities of Bridgeport, Hartford, and New Haven, Connecticut, in the spring of 1909. The previous year's legislative effort to tighten child labor regulations in the Nutmeg State had failed, "due either to weakness of our forces, or to earnestness and strength of the opposition," as Owen Lovejoy, general secretary of the National Child Labor Committee (NCLC), explained forthrightly.[4] Lovejoy sent Hine into the state in March

1909 to gather evidence intended to sway public sympathies in favor of reform. On this initial foray into New England, Hine photographed newsboys and newsgirls, messengers, a couple of bowling alley boys, and one bootblack. He also photographed individuals and institutions in the region engaged in efforts to change children's presence in city streets. Thus his photographs document the threats posed to children by the streets while also pointing the way toward the resolution of those threats by depicting the spaces of reform. Hine's images aided both local and national campaigns to regulate the lives of children in city streets where they worked and played. Hine was doing his part not only to reform child labor but also to remake the urban landscape with more hospitable opportunities for children.

While the city street was widely condemned in popular culture, the figure of the newsie was often highly romanticized. In her memoir *The House on Henry Street* (1915), Lillian Wald, the settlement house director and public health advocate who served on the board of the NCLC, noted that "opposition to regulating and limiting the sale of papers by little boys on the streets is hard to overcome." She attributed this resistance to "juvenile literature" that "glorified the newsboy and his improbable financial and social achievements."[5] Interest in the newsboy was "heightened by a series of pictures by a popular painter, wherein ragged youngsters of an extraordinary cleanliness of face were portrayed as newsboys and bootblacks." It became Lewis Hine's job to offer a counter-narrative. As Wald stated, opposing "the charm of this presentation" was the "practical reformer [who] offers the photographs, taken at midnight, of tiny lads asleep on gratings in front of newspaper offices, waiting for the early editions." Rather than romance, this reformer "finds in street work the most fruitful source of juvenile delinquency."[6] Hine was charged with capturing images of the degraded and debased newsie, and he pointed his lens in that direction.

Social critics such as Edward Clopper, the secretary of the National Child Labor Committee for the Mississippi Valley, cautioned against reformers' tendencies to cast all street activity as unwholesome, to portray the street as "a black monster" with "nothing but evil in it." Clopper, who wrote extensively on the street trades, reminded his readers of the importance of seeing how closely the street was "woven into the life of every city dweller, for his contact with it is daily and continuous. If it is all evil," he argued, "it ought to be abolished; as this is impossible, we must study it to see what it really is and what needs to be done with it." Clopper recognized the vital civic role streets play in bringing people into "closer touch with one another," creating spaces "where they meet and converse, where they pass in transit, where they rub elbows with all the elements making up their little world, where they absorb the principles of democracy,—for the street is the great leveler."[7] Nonetheless, Clopper warned that the street "was never intended for a playground, nor a field for child labor, nor a resort for idlers, nor a depository for garbage, nor a place for beggars to mulct [swindle] the public. These fungous growths from civil neglect must be cut away."[8] Hine proceeded to fashion sympathy for reform of the street and of the young workers who toiled there.

## HINE PHOTOGRAPHS NEWSIES

The first image in Hine's Connecticut series (fig. 2.1) is an example of the elaborate cultural staging Hine used to align his images of newsies with popular anxieties over the unsupervised life in "the street." The photograph shows two boys, each with newspapers under his arm, crouching across from each other on the sidewalk and broad stoop in front of a doorway. A sign behind them advertises "Hot Tom and Jerry," a popular seasonal drink of eggnog mixed with brandy and rum. One boy, in full silhouette, has a smile unfolding and a hand open and reaching out. His companion's back is to the camera, and we see only the edge of his modest bundle of papers as some interaction takes place. Both boys are wearing caps and jackets showing use but still suggesting warmth, and their shoes appear worn but serviceable. The boys are situated in the center of the somewhat dark gray image. They blend into the urban landscape; thus the photograph joins their presence with reformers' anxieties.

Hine's caption elaborates on and thereby alters the photograph's meaning, pushing it in the direction of his professional purpose: to arouse sentiment favorable to the passage of child labor legislation. The image itself offered a modest challenge to the well-ordered society. The boys appear nondomesticated and unsupervised. From Hine's caption, however, we learn that the boys, whose names we don't know, were brothers, eight and ten years old, and had been selling papers in Bridgeport, Connecticut, for two years. The older boy told Hine that his younger brother, working between three in the afternoon and nine or later at night, earned as much as fifty cents a day. But in this photograph, the boys are not selling the papers bundled under their arms. It's after nine, and they have pitched pennies "all evening." Do the boys' parents know where their children are? If they do know, we are left to question their parenting skills. Hine reports seeing the younger brother going into the local saloons "with a big bunch of papers and a pitiful tale." He leaves to his viewers' imaginations the heartrending stories that might envelop the boys, who for now reside only within the picture frame. Gambling, saloons, low wages, and long hours were threads that wove a portrait of moral transgression and parental abandonment and portended societal danger.

Set within the context of the reform movement, images such as this implicitly raised the question of whether or not the state should assume parental authority and, through restrictive labor legislation, discipline the behavior of its young wards. Hine suggested that the boys needed some form of parental constraint. In the introduction to his popular book (1915), in which many of these images appeared, the reform

FIGURE 2.1. "9:00 P.M. Bridgeport, Conn." (Hine 0587)

advocate Philip Davis of the Civic Service House in Boston located the threat more directly, arguing that street influences were "undoing the work of home and school." Davis believed that Hine's "Servants of the Common Good" should be trustees of the street and determine its appropriate uses, because "street children, like the streets, are in a peculiar sense, public property of which the community is the trustee."[9] If the children who labored in the streets were "public property," as Davis argued, they were abandoned property in need of claiming. Reformers such as Davis sought to expand the public's civil ownership of these young workers and its sense of responsibility for their protection.

Hine's photographs repeatedly provide dramatic evidence to support arguments for a more interventionist state, given the apparent absence of parental oversight and appropriate familial protections they depict. Hine intended the image in figure 2.2 and its caption to be photographic theater:

> 9:30 P.M. A common case of "team work." Smaller boy (Joseph Bishop) goes into saloons and sells his last papers. Then [he] comes out and his brother gives him

more. Joseph said, "Drunks are me best customers." "I sell more'n me brudder does." "Dey buy me out so I kin go home." He sells every afternoon and night. Extra late Saturda[y. At] it again at 6 A.M. Sunday, Hartford, Conn.

Hine packs several themes into this caption. Hine often photographed family members joining in a shared work experience, older brothers and sisters initiating their younger siblings into the trade. In this instance, Hine notes the "team work" between the brothers, but only to suggest that the older one, Meyer, who was twelve when the photograph was taken, was innocently leading his eleven-year-old brother Joseph down a path of misery and immorality. Hine posed the boys about to descend into the lubricious world of saloons and cafés, urban places where liquor was served and no child should ever be found. Hine puts words into Joseph's mouth, words that Hine translates into class signifiers such as "Dey," "kin" and "brudder," and which are leavened with the slight hint of a brogue when Joseph confesses that "drunks are me best customers." Hine is telling a story about childhood lost to long hours spent in the wrong places with the wrong people and in pursuit of the wrong ends, all for a half-dollar payday. The family suggested by this image is socially incomplete—not a haven in a heartless world but an inadequate model of the ideal family that Hine and his fellow reformers used for comparison.

We could read figure 2.2 with another understanding of the street and of family. The historian Laura Wexler argues that "for photographs to communicate, the viewer must in turn be able to read and interpret them, like other languages and signs." Meaning is partly derived, she writes, by bringing the image "into relation to other, publicly legible, semantic structures—myth, ideology, semiotic systems. To be seen, photographs must be woven into other languages." Otherwise, they are prone to "float off in an anarchy of unincorporated data."[10] If we read this image from the cultural perspective of the children and their families, it would seem only natural that siblings work together and support and protect one another, with older brothers looking out for younger brothers even while putting them in danger. The Bishop brothers knew the virtues and responsibilities of brotherhood, each brother having strengths and weaknesses that the two of them had no problem using to their advantage. They recognized that in the business of selling newspapers, youth sells. They understood the drunk's vulnerability, if not their own. They entered saloons to sell, and that's what they did, until they cashed out for the day. Exactly what they needed the money for was not evident. Joseph and Meyer were the first and second sons of Jack and Ethel, Yiddish-speaking Jews who had arrived in the United States from England only a few years earlier, in 1907. The father was a fruit peddler, suggesting that he was familiar with life in the streets. Ethel showed no occupation, though having borne six children in rapid succession, she clearly was laboring in more ways than one.[11] Work was the natural

FIGURE 2.2. "9:30 P.M. A common case of 'team work.'" (Hine 0595)

condition among the Bishops and their neighbors, no matter their age. So the subjects of this image and their family were not an aberration presented to Hine's viewers but rather were very much a part of the world of work and the urban fabric that reformers condemned.

Hine relied on visual props, when they were available, to suggest impropriety. For instance, one of Hine's images (fig. 2.3) is of a ten-year-old girl in Bridgeport who "peddles papers." She had been "at it" for one year. Her presence in the streets alone would have been a matter of concern and caution among reformers. This image, a frontal portrait set within the subject's place of work, is typical in Hine's archive. The young girl appears to be well dressed, wearing a hat and ribbons in her hair, seemingly warm boots, and a thick-looking wool coat. We don't know her name, but she was representative of young girls on the city streets. She is smiling widely, but by the way the image is structured, with the girl standing in front of a sign pitching "Lager Beer, Ales & Porter," the viewer might guess that her cheerfulness is a sign of naïveté about the city's dangers. Hine mixed innocence with an urban landscape littered with temptations and moral challenges.

In another of Hine's didactic images, labeled "Early Education" (fig. 2.4), Hine

FIGURE 2.3. "10 year old girl who peddles papers." (Hine 0590)

portrays a twelve-year-old boy sitting on a stoop while reading a newspaper. We might applaud his reading if he were in an appropriate place—at home or the library, or in school while being taught by a qualified teacher with the best interests of the child in mind. Hine is suggesting that children in the street trades are in fact gaining an education, leaving open the question of what they are learning and from whom. With adult men standing on the sidewalk on either side of the boy, possibly indifferent to and unaware of his existence, Hine hints at an unfortunate answer. For his part, the boy appears equally oblivious to what is around him, seemingly content to be where he is, his attention riveted on his reading. With the street as his school, his education, many reformers believed, could lead in one direction only—to his moral decline.[12]

Hine's images show children who could navigate the city despite the apparent dangers—children who were accustomed to moving unnoticed through the streets, factory lots, back yards, and refuse dumps, along train and trolley tracks, through licit and illicit terrains. In none of the photographs do children look frightened or intimidated by their surroundings. One of Hine's Hartford photographs (fig. 2.5) is a group portrait of four girls and three boys of various ages—smiling, playful-looking children (such as are present in other of his photographs). His caption identifies only what

FIGURE 2.4. "Early Education." (Hine 0593)

reformers might consider the more transgressive behavior—"Girls coming through the alley"—and informs the viewer that even the smallest girl in the image is an experienced newsie, with two years in the trade.[13]

Unlike newsboys, newsgirls did not appear in large groups. Hine more frequently photographed newsgirls as individuals or in small groups of two or three. He did once place newsgirls within a large group of newsboys even though no girls were physically present, captioning a photograph of a group of newsboys taken at the *Times* office in Hartford (fig. 2.6) with the single remark that the "newsgirls mingle and jostle these boys in the office and in the yard." The image is full of young boys, the youngest in front ascending to older boys in the rear; yet Hine provided a caption whose subjects are out of the frame. Hine implicates the newsgirls in the rough-and-tumble world of the newsboys, thereby playing to a powerful social anxiety.

An image more caustic to reformers' professional sensibilities was that of the Hartford newsgirl Mery Horn, "a hunchback" (fig. 2.7). Mery is dressed in a thick wool coat and has a warm hat on her head. With a slight smile on her face, Mery looks hopeful, innocent. Off to one side is a woman, a sliver of her dress framing the right border, her gloved hand perhaps straightening her hat; her bare wrist is showing, and

FIGURE 2.5. "Girls coming through the alley." (Hine 0594)

FIGURE 2.6. "Group of newsboys at the Times Office." (Hine 0616)

her elbow is at a right angle to Mery, who stands at the corner of the building. Hine captioned the image with a statement attributed to Mery's doctor—Mery is a "severe case" whose ill health is "being aggravated by the heavy load of papers" she carries. Ordered to stop, she replies emphatically to the contrary, invoking her interests in the matter: "But I need the spending money. I have to go to shows." How much of this conversation Hine was actually a party to we don't know. His photo and caption portray Mery as a victim of ill health resulting from or exacerbated by her work as a news seller. Mery didn't share the perspective of reformers, however; her reality was best expressed by her insisting on her own moral values and opting for more shows. Work provided Mery with access to the world of popular entertainment and commercial amusements, a world that the historian Kathy Peiss and others have shown was nearly irresistible to young women and men of that era.[14]

If the children in Hine's photographs did not seem particularly intimidated by the urban environment, Hine did record situations that cried out for intervention and children's protection. In his image of Tony Casale (fig. 2.8), an eleven-year-old newsboy in Hartford, Hine gives us a glimpse of the pain that some children endure on the road to adulthood. Hine was generally able to convey the dignity of the children he

FIGURE 2.7. "Mery Horn, a hunchback." (Hine 0591)

photographed, but within Tony's face we see a troubled look that leads us to imagine the indignities Tony lived with, an interpretation unavoidable in light of Hine's caption for another picture he took of Tony. In that caption Hine confides that Tony had been selling papers since he was seven years old, sometimes until ten at night, and said he met "Drunken men [who] say bad words to us." Although such information was certainly fodder for the reform mill, it paled beside Hine's report of the abuse Tony received at home. Again Hine was not satisfied merely to photograph and depart: he spoke with Tony's paper boss (and probably with others), who reported that Tony had shown him "marks on his arm where his father had *bitten* him [Hine's emphasis] for not selling more papers."[15] Hine's caption is essential for that detail, revealing the violent domestic world Tony lived in.

Newsies' association with street life put them in great social danger. In an article in the philanthropic journal *Charities and the Commons,* the economist and social activist Scott Nearing, then a professor at the Wharton School of Business at the University of Pennsylvania, wrote: "Whatever the cause, the effect on the newsboy is always the same. He lives on the streets at night in an atmosphere of crime and criminals, and he takes in vice and evil with the air he breathes. If he grows into manhood and escapes the tuberculosis which seizes so many of these boys of the street, the things that he has learned as a professional newsboy lead in one direction,—toward crime and things criminal. The professional newsboy is the embryo criminal."[16]

Suggestions of "the embryo criminal," of children fallen to street life, of children already succumbing to antisocial behavior, imbued Hine's photographs with the empathic power he sought. Very rarely did Hine display the children's self-destruction in his images alone; his captions were critical to bringing out those revelations. In Hartford, for instance, Hine photographed a group of six boys (fig. 2.9), but his caption names only one of them: Joseph De Lucco, the "boy in middle" and an eight-year veteran of selling newspapers. If the caption had left off there, we might look at this group and see some enterprising young men and a backbench of young followers. However, Hine goes on to inform the viewer that Joseph "was arrested for stealing papers a while ago." Joseph's eight years in the street, we are to assume, had already left an indelible mark on him.

Reformers and community leaders alike put newsgirls under even greater scrutiny for any hint of impropriety. Hine often contextualized his work within these social anxieties. Take the image of Hartford newsgirls Alice and Besie Goldman and Bessie Brownstein as an example (fig. 2.10). Once again, Hine's photograph alone does not hint of transgressions. At a glance, the image has three young girls in the central focus and an even younger boy on the left side, set slightly back from the girls and standing inside a doorway. They all look well-enough dressed, with warm coats and hats. According to Hine, the girls are waiting for their papers. He identifies the three: Besie

FIGURE 2.8. "'Bologna' See also photo 600 and 655. Hartford, Conn." (Hine 0660)

FIGURE 2.9. "Hartford, Conn., newsboys." (Hine 0613)

and Bessie are both nine years old and have been selling papers for one year; Alice, ten, has been selling for four years. That the three were working daily until 7:00 or 7:30 in the evening is a matter of concern. Hine packs the photograph with greater emotional power by including in the caption the harsh condemnation of a newsdealer who told Hine that Alice "uses viler language than the newsboys do."[17] The girls' imminent downfall is not yet etched into their faces, so Hine emotionally charges the image by making that connection for us. No matter the circumstance, female children working in the streets as newsies faced much higher moral standards; their misbehavior demanded greater consequences. While the children may be faultless, those seen as promiscuous young women often carried that stigma to the reformatory or to jail.[18]

## HINE PHOTOGRAPHS MESSENGERS

If newsies were susceptible to debilitating habits, telegraph messenger boys were even more endangered by their work. Not only did the messenger boy work the streets, often at night, but he was "frequently found in the worst resorts of the 'red-light' districts of our cities." Reformers considered messenger service, like other street occupations, a blind alley leading nowhere. U.S. Commissioner of Labor Charles P. Neill noted that while "the newsboys' service is demoralizing, . . . the messenger service is debauching." The saddest of all, Neill confessed, was that messenger "service appeals strongly to the children."[19] John Spargo, the best-selling socialist author of *The Bitter Cry of the Children* (1906), bemoaned the fact that the messenger boys liked working in New York's vice-ridden Tenderloin district, not because they were "bad or depraved" but because of the economic incentives of such work.[20]

Spargo was wrong: money was not the only incentive that made messenger service popular among young boys. Messengers, like newsies, were particularly susceptible to romanticization in the public mind. Dressed in their uniforms, dashing about the city, the messenger boys projected a certain élan. If Hine's job was, in part, to remove the romanticization that often engulfed the boys, then it is safe to say that Hine sometimes failed in his work. An excellent example is his photograph (fig. 2.11) of a Western Union messenger in Providence, Rhode Island, who was known locally as "Speed," a moniker no doubt earned by his proficiency in the trade. Hine positions the boy in the center of the image, with the gutter running into the horizon, an automobile in the distance (perhaps suggesting the future), and a horse-drawn wagon (suggesting the past) cut off on the left edge of the composition. The boy is decked out in his uniform with its double-breasted jacket, and his cap is tipped to one side as a matter of style, as he leans against a telephone pole in a show of comfortable cool. The uniform transforms the boy's appearance and alters and elevates his place in the neighborhood. Here, as in some of his other photographs, it was difficult for Hine to

FIGURE 2.10. "Newsgirls waiting for papers." (Hine 0597)

FIGURE 2.11. "'Speed.'" (Hine 3169)

lessen the attraction of a boy in uniform. To rescue messengers from their fate, Hine situated his appeal for sympathy around questions of fundamental concern to the public—questions about when and where the messengers worked and whom they encountered on the job.

Hine was occasionally able to destabilize the messenger boy's dashing, romanticized figure, photographing messengers in an act of transgression and thus undermining the image of the innocent boy in uniform. In these photographs, the threat is easy to discern, even to the untrained eye; the action is sufficient to articulate a message of concern for the messenger boys' fate, without the assistance of a caption. In one image (fig. 2.12), Hine captures the boys in an illicit activity, but he then supplements the image with the details of their ruin, once again to amplify the warning that boys left unsupervised pose a great danger to their own welfare and perhaps to the interests of their families. Hine's photograph takes the viewer into the "Den of the Terrible Nine," the waiting room for Western Union messengers in Hartford, where the boys are "absorbed in their usual Poker game." Hine confirms our suspicion that they are gambling, informing us that they "play for money." More than a simple game between boys, according to Hine's caption ("Some loose [sic] a whole month's wages in a day and then are afraid to go home"), this is a real vice with tangible losses or gains. Even before they leave to deliver their messages, the boys place themselves in danger, and there are even greater threats awaiting them out on the streets.

Messenger boys were particularly susceptible to the ravages of vice and immoral temptation because their work often took them to the worst sections of the city. Drawn into this web of vice by the promise of greater financial rewards, the young messenger boys showed little initiative to depart the trade. In these "moral cesspools," children were called on "for many services by the habitués of these haunts of the vicious and the profligate": they might be "sent out to place bets; . . . to take notes to and from houses of ill-fame; . . . or to buy liquor, cigarettes, candy, and even gloves, shoes, corsets, and any other articles of wearing apparel" for the working girls. It was not unusual for messenger boys to "know many of the prostitutes . . . by name" and, in many instances, to proffer personal recommendations.[21]

According to Edward Clopper, messengers' work changed dramatically around nine or ten o'clock at night, after which time delivering telegrams was only a small part of their job. Clients sent messengers on a host of errands that essentially put them at the service of the underworld. The uniform and cap of the messenger acted as "a badge of secrecy," enabling him to "get liquor at illegal hours or to procure opium and other drugs where plain citizens would be refused." Nightly association with "the lowest kind" left the messenger "to regard these evil conditions as normal phases of life." Prostitutes had their favorites, "usually the brightest boys." Clopper noted that the prostitutes "take a fancy to particular boys because of their personal attractiveness

FIGURE 2.12. "8:00 P.M. Flashlight photo of messengers absorbed in their usual Poker game in the 'Den of the Terrible Nine.'" (Hine 0592)

and show them many favors, so that the most promising boys in this work are the ones most liable to suffer complete moral degradation."[22]

Hine often photographed messengers whose moral destiny was in part governed by the references in Hine's captions to the children's debilitating affinity to places and people of vice. In one image made in New Haven, for instance, Hine shows a group of messenger boys going home after a shift (fig. 2.13). The meaning of the image is transformed by the caption, which alerts viewers that the telegraph offices where the boys work are "almost next door to a caf[é]—boulevard frequented by street walkers." Hine observed, "Many of [the] women parade the streets and the boys meet them constan[t]ly and are called frequently into house[s] of ill repute." For Florence Kelley and many of her contemporaries, it was simple: messengers had unfortunate "contact with disreputable people," not just in the street but also at the numerous addresses they visited each day, "places of the existence of which more fortunate children are carefully kept in ignorance." A judge who presided over a "justly famous juvenile court" stated his belief, according to Kelley, that "two-thirds of the messages delivered after eight o'clock at night in his city were carried by children to places of bad character."[23] After dark, the city abounded with people and places no child should visit. To emphasize that fact, Hine took several photographs of messengers at night and alluded in his captions to the likely threat from gambling, vice, and immorality.

## NEWSBOYS' REPUBLIC

Advocates of clearing the streets of child laborers had only limited regulatory power to address the issue. With regard to messengers, little remained but to raise the age limit for messenger boys to eighteen and prohibit night work. As for newsies, the younger of the two groups, reformers strove to regulate the newspaper-selling trade and provide the children with alternative spaces. Lillian Wald advocated a system like the one in Boston, which required the consent of parents before a boy could obtain a badge from the district school superintendent granting permission to sell papers on the street.[24]

Some cities took modest steps to protect children who labored in the street trades. Among American cities, Boston had one of the most developed programs to reduce delinquency and imbue a sense of personal responsibility in newsboys. Boston, joined by New York and Buffalo, led national efforts to restrict the very young from the newspaper trade. In all three cities, legislation prohibited newsboys under the age of ten from plying their trade. Whatever the limitations of the legislation, the "prohibition of street work for children under the age of ten years registers a distinct ethical gain," Florence Kelley insisted.[25] For children aged ten and older, Boston school authorities regulated the newspaper-selling trade.

In 1909, approximately three hundred newsboys in Boston were brought before the juvenile court for violation of the city's license laws. To avoid clogging the judicial docket, authorities were determined to establish a "newsboys' court" to govern activity in the trade. The following year, a petition to establish such a court was presented to the Boston School Committee, which approved the idea. On election day, newsies cast their votes for three juvenile judges of the court, and these judges, together with two adults appointed by the school committee, made up the newsboys' court called the Trial Board. The court was empowered to investigate and report alleged infractions to the school committee. Newsboys could bring their colleagues before the court to account for numerous violations, such as failure to wear a badge while selling, working after eight o'clock or on streetcars, bad conduct, irregular school attendance, and gambling or smoking. Disposition of these cases varied from reprimands and probation to the suspension or permanent revocation of newspaper-selling licenses.[26]

Boston supplemented efforts at regulating street trading by organizing a "Newsboys' Republic." From the newsboy's court emerged a plan for "self government among the licensed newsboys through the so-called Boston School Newsboys' Association." The

FIGURE 2.13. "11 P.M. Messenger boys going home." (Hine 0596)

association pledged itself to "the enforcement of the license rules and the suppression of smoking, gambling and other street vices, more or less common among the street boys of certain neighborhoods." The newsboys ran the association, selecting officers of their own choosing—one newsboy captain and two lieutenants for each school district, as well as a chief captain and general secretary and an executive board consisting of seven newsboys elected from the ranks of the captains. Their goal was to see that license rules were adhered to and that the principles of the association were followed in each school district. Association officers conducted weekly inspections on the street, supplemented by monthly inspections at the schools.[27]

Hine attempted to record the Boston experiment. He photographed the Newsboys Building in 1909 (fig. 2.14), and in the caption for the image he noted that the building was "intended to counteract the attractions of the street." It was a central institution in the constellation of institutions that emerged in the late nineteenth and early twentieth centuries and that were created with the purpose of providing uplifting alternatives to the street's debilitating attractions. The Newsboys Building was an urban place where boys could play, take part in uplifting activities, govern themselves, adjudicate disputes, and administer justice, all while forging community with their fellow toilers.

FIGURE 2.14. "The Newsboys Building." (Hine 0952)

Florence Kelley acknowledged progress. "There have been newsboys' homes, lodg-ing-houses, banks, and clubs; newsboys' picnics, public dinners, treats and even, from time to time, a theatrical performance for newsboys." Kelley applauded the fact that the "simple device of prohibiting the work of tiny children and making the privilege of selling papers on the streets, out of school hours, depend upon the good behavior and regular attendance of the candidate at school, registers a marked gain in reasonable kindness of the communities which have entered upon this humane course of action." The array of institutions addressing the needs of newsboys surpassed the number of institutions of uplift for all other types of child labor.

Communities such as Boston that required children to wear badges indicating their right to sell papers provided a model for other communities. Wearing a badge gave a clear signal to all that such children were "school boys authorized by their parents and the board of education to work, out of school hours, until ten o'clock at night." Under such a system, every child wearing a badge was vouched for by a parent or guardian and was known to local educational authorities. "There are no waifs or strays among them," Kelley proclaimed. Furthermore, "They are not legitimate objects of pity or of charity. They are school boys in good standing." The next step, she contended, was

FIGURE 2.15. "4 p.m. Morrison Foster." (Hine 4644)

to raise the minimum age of street workers to twelve and to restrict the workday for children under the age of fourteen to the hours between seven in the morning and seven at night.[28]

When photographing newsies in Boston, Hine kept a lookout for the badges. Hine was not reporting missing badges but was assessing the success of the Boston system based on his own observations derived from survey in the field. Hine photographed a group of newsies in front of Boston's South Station; four of the boys confessed to being eleven years of age, but Hine "saw no badges in evidence."[29] Also at South Station, at four in the afternoon, Hine photographed twelve-year-old Morrison Foster, who told Hine that he sold papers from four o'clock to six o'clock each day (fig. 2.15). Not seeing a badge in evidence, Hine questioned Morrison, who showed that his was pinned inside his coat pocket. Many newsboys did this to protect their badges from being stolen. South Station was a very busy place, and Morrison, holding a large bundle of papers under his arm in the photograph, sold quite a few in the two hours of after-school peddling.

## THE STATE, CIVIL SOCIETY, AND THE STREETS

Hine's work provided a conduit between the concerns and campaigns of the National Child Labor Committee and those of local community representatives. Both parties were engaged in defining issues of child welfare, but they applied resources and strategies from different perspectives. Critically, local individuals and organizations opposed to child labor often led Hine to his photographic subjects; he did not always just stumble upon them. This is not to say that he was unable to find his subjects without assistance; when he photographed textile workers, for instance, he located his child subjects easily by observing known textile mills. To get inside a factory, Hine deployed many ruses. According to the historian Alan Trachtenberg, Hine had a reputation as a good actor. In his work for the NCLC, his "repertory ranged from fire inspector, post card vendor, and Bible salesman to broken-down schoolteacher selling insurance. Sometimes he was an industrial photographer making a record of factory machinery."[30] Photographing in the streets was different, however. Without local knowledge, Hine found it difficult or time consuming to find likely targets for his camera to record. In those instances he benefited from the advice and direction of local experts on local conditions.

Although we don't have records from Hine's 1909 Connecticut survey, we do have Hine's handwritten notes scribbled on the back of hotel stationery from his 1916 investigation of the street trades in Vermont. These notes help reconstruct Hine's methodological approach to survey work, indicating that Hine did far more than simply wait for children to appear before him so he could snap a photograph. He went looking for local opponents of child labor and anyone who had regular contact with children, from teachers and school superintendents to social workers, settlement house activists, juvenile justices, and labor inspectors. Hine asked questions, followed up leads, and inquired about specific conditions in Burlington.[31] In Connecticut, Mary Graham Jones, a former librarian turned settlement house activist, pointed the way for Hine and provided him with additional opportunities to portray the positive work being done in that state to address child labor and the street trades.

Mary Graham Jones became head social worker at the Hartford Social Settlement in 1900 and devoted her last years to settlement work in the city. In *Woman's Work in Municipalities,* Mary Ritter Beard lauded Jones for all she did "during her life for the betterment of child life and neighborhood life in her native city." Beard considered Jones an "inspiration to other public-spirited women."[32] Perhaps Jones's greatest legacy to child welfare in Hartford came a few weeks prior to her death in 1915, when she submitted a plan to city authorities for the establishment of small local playgrounds for young children in various neighborhoods of the city. Jones believed that children needed an alternative to the street. She devised a scheme to situate playgrounds in

FIGURE 2.16. "Miss M[a]ry Graham Jones of the Social Settlement." (Hine 0601)

local neighborhoods so that they would be near enough and safe enough for young children to use.

Jones's effort to secure playgrounds for Hartford children illustrates how the state developed and expanded its social welfare responsibilities. Jones proposed leasing a dozen or so vacant lots from the city, "at a nominal rent," and then having the city's parks department develop the lots into functioning playgrounds, with the playgrounds subsequently to be supervised by the department of education. City aldermen approved the project and granted $2,500 for the first year's expenses; nearly all of that money went to the parks department to create the small playgrounds, with "various successful results." The city took up maintenance and support of the new citywide playground system and almost immediately made plans to expand activities to include provisions for winter sports such as skating. In honor of the plan's sponsor, a playground set aside for children under the age of nine was named the Mary Graham Jones Playground by the citizens of Hartford.[33]

Hine photographed Jones and a "few of the newsboys (all 10 years old and younger) whom she invited to a Sunday afternoon party" (fig. 2.16). Jones and her enthusiastic flock are positioned front and center in the image, below a railroad yard that divides the image horizontally. A man overlooks the scene from the upper level behind Jones and the children. He is apart from, yet also a part of, this group. He appears as an authority figure of sorts, looking down on the assembly. Perhaps we should read him as the state overseeing and intervening to support the work of Jones and the settlement house workers. Hine's caption also reveals that in two days Jones secured the names of thirty-nine newsboys and newsgirls under the age of eleven. She was settled in the community, reaching out to the children of the city and attending to their real and perceived needs.

Lewis Hine not only photographed what was wrong and needed fixing; he also photographed what was right, what reformers were doing to protect children and forestall their further degradation. In New Haven, Hine took photographs inside two of the most successful local institutions for keeping children off the streets of the city—the United Workers Boys' Club and the Bancroft Foote Boys' Club. "Yale men" were involved in the founding and direction of both clubs. In the first twenty years of the United Workers Boys' Club existence, its superintendents came from Yale, and work in the club strongly appealed to undergraduates at the university. Likewise, Yale students and faculty took great interest in the Bancroft Foote Boys' Club. The club had strong ties with the Yale Young Men's Christian Association, which "encouraged and aided all these activities of the students, and broadened them." According to one historian of the city, "the leaders of Yale were living the life of the city. They were making its problems their own." Yale faculty and students also played an important role in the Lowell House Social Settlement, an institution in whose progress "Yale idealists had from the first a definite part."[34]

Hine's photographs of the United Workers Boys' Club and the Bancroft Foote Boys' Club portray the virtuous ideal sought by reformers and the anticipated advantage these institutions had over life in the street—they introduced order into the boys' lives. Philip Davis applauded the "many brilliant experiments for taking children off the streets, which have inspired the founding of many a social settlement, boys' club, vacation school and recreation center" and called for "further experiments to organize and supervise the lives of the many thousands of children still on the streets and destined to remain there for some time."[35] In the caption to his photograph of the Bancroft Foote Boys' Club (fig. 2.17), Hine notes the "common scene" of boys shooting pool, describing it as "one way to control the street boys." The boys are not in a pool hall, mixing with people of questionable repute and gambling away their day's earnings. The boys are under adult supervision and out of trouble. Hine produced a similar image at the United Workers Boys' Club (fig. 2.18), and again the emphasis is on well-ordered, supervised play.

Hine also showed who benefited by these services. He photographed twelve-year-old Hyman Alpert, for instance, who told Hine he had been selling papers for three years. In his image of Alpert, Hine is able to contrast Alpert's work life—where he was "Bought" and "Sold," according to the advertising copy on either side of him—with his life in the boys' club, where he spent his evenings and could simply be a boy.[36] Hine

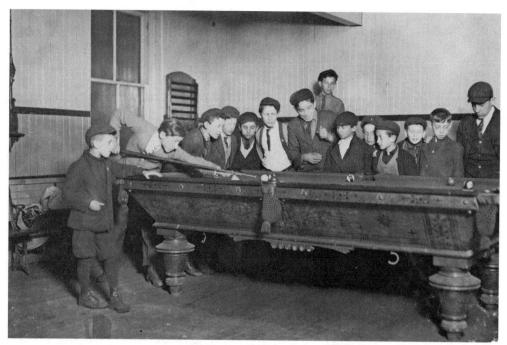

FIGURE 2.17. "One way to control the street boys." (Hine 0649)

FIGURE 2.18. "United Workers Boy's Club. New Haven, Conn." (Hine 0614)

FIGURE 2.19. "United Workers Boy's Club, New Haven, Conn." (Hine 0615)

also highlights the health benefits of physical activity in his photograph of boys doing calisthenics in the gymnasium at the United Workers Boys' Club (fig. 2.19). Through organized recreation, the boys' bodies had an opportunity to grow strong.

The children's needs were obvious, and the urge to help universal, according to Philip Davis, but what was required was an "enlightened" view of children. In the "new view of the tender child," Davis argued, the child "is incapable of crime and, whether in or out of mischief, is always in need of protection, encouragement and care." With new juvenile courts, the state might be sympathetic, but it was not to be apathetic. "Street activities," Davis declared, "must henceforth be organized under direction and close supervision." With this new view of childhood, it was imperative to see that children became full and active citizens before the "embryonic criminal" grew into maturity.

In play, as in reform efforts, organization was critical. Davis hoped to enlist the public in the effort to protect children. Davis acknowledged that the call for help had engaged the assistance of "business men, club women, physicians, lawyers, ministers and editors, as well as plain fathers and mothers." This was vital assistance, though Davis noted that it was being carried out in "a somewhat disorganized fashion." He believed that every neighborhood should serve as a "searchlight on its own street conditions in order to locate and eradicate the destructive influences and to lift the life of the street to the level of that which is best in the life of the home and school." Davis was also staking out new ground for the state when he argued that "street children, like the streets, are in a peculiar sense, public property of which the community is trustee." Urban education and recreation departments needed to represent "the community's sense of obligation" to children, much as the street department does "toward public property of one kind."[37] Sympathy for the child necessitated an expanded state. As a generator of public sympathy, Lewis W. Hine's benevolent surveillance provided the evidence for local change and the rationale for policies promoting the greater involvement of the state in the lives of children.

## CHAPTER 3

# Textiles

———

I N THE FIELD OF CHILD LABOR reform, no industry received more attention than the textile industry, and deservedly so, because no other industry could match its ability to put children to work. American children were central actors in the industrial workforce from its birth in 1790. On that cold December morning in Pawtucket, Rhode Island, at a fulling mill on the Blackstone River, when Samuel Slater set the waterwheels moving and the gears turning to transfer power to specialized machines designed to spin cotton thread, he did so with a workforce made up of nine children between the ages of seven and twelve.[1] High numbers of laboring children remained a central feature of the textile industry into the twentieth century. But by the time Lewis Hine came to New England in 1909 to photograph children at work, the geography of the textile industry was changing. In the 1870s, capital in the industry began to shift from New England to the Carolinas, Georgia, and Alabama, and child labor followed suit. In fact, the availability of cheap labor—that is, child labor—was one of the inducements for textile companies to transfer capital. Competition between North and South extended not only to spindles but also to the efficacy of child labor laws.[2]

## CHILD LABOR AND REGIONALISM

When Hine began photographing children at work in textile mills, he did so within the context of a raging sectional debate over the intentions of child labor reformers and the responsibilities of textile manufacturers. Child labor followed the textile industry wherever it went, so when the industry moved from New England to the South after Reconstruction, child labor became deeply rooted in southern industry. A. J.

McKelway, the National Child Labor Committee's secretary for the southern states, told attendees at the committee's annual conference in 1910 that "the Southern manufacturer simply copied the system he found in New England and bought machinery adapted to child labor just as the New England manufacturer had done when he copied the system and machinery of Old England."[3] In McKelway's view, New Englanders played a considerable if indirect role in bringing child labor into the South.

Some observers took the accusations further, noting that many northern textile companies retained a controlling interest in southern plants that operated with large numbers of child laborers. Rabbi Stephen S. Wise of New York City's Freedom Synagogue stated, and others agreed, that the "capital-lacking Southland" succumbed to "the tempting and oppressing capital of the North." Having "built child labor mills in the South," northern capital was guilty of "long-distance or wireless sin—for the northern capitalist gets his dividends from southern child-labor products, though divided by a thousand miles and more from his little victims." Northern capital was not only "guilty of southern child labor" but also of "impoverishing and damning the South's future" and "unconsciously taking its revenge in crushing out the lives of the little children of the Southern States." Wise argued that child labor might be "under control" in New England, but "northern capital is not under control."[4]

New England manufacturers naturally denied culpability. J. Howard Nichols, treasurer of the northern-owned Alabama City Mill, countered the charges against northern capital by responding that there was "nothing within . . . or without" the Alabama City Mill "and the little town of 2300 people which has grown up around it" of which "any citizen of Massachusetts need be ashamed." On the contrary, he argued, the mill's owners transplanted into the region one of New England's most treasured exports, the New England village, establishing "a model town that should be an object lesson at the South," complete with a church building that "would be an ornament in any village of New England."[5]

There was a growing conviction among child labor reformers that, despite real differences between the North and the South, children in those regions and in all other areas of the United States should have the same rights and protections. Unfortunately, the contrary was true. The incremental state-by-state nature of child labor reform created unequal protection for children living in different parts of the country. As Samuel McCune Lindsay, a professor of social legislation at Columbia University, stated, "unequal laws for the protection of childhood in the several states of the Union" also meant "unequal standard[s] of civilization." The fundamental question was, "is it fair that . . . our state legislation . . . should make the birthright of the American child mean less in one state than in another?"[6] Child labor was controlled by diverse regimes of laws that differed from state to state. Some states such as Massachusetts that had strong protections for children also had high levels of child labor.

Racial concerns united the northern and southern regions behind child labor legislation and compulsory school requirements. Southerners recognized that white supremacy depended on more than just an assertion of superiority. Proponents of compulsory education argued that since blacks were largely denied mill work, they were in school while white children toiled in the mills. Lewis Parker, a cotton manufacturer in Greenville, South Carolina, testified to Congress: "I recognize the fact that the Negro believes that his salvation lies in education, and we have not yet impressed upon our white people the fact that his superiority must continue only through intelligence and through superior education."[7] Rather than white children being released from the mills to attend school, however, black schools were starved for resources, and thus all children suffered.

In New England, racial fears were focused not on blacks but on immigrants and their foreign cultures. In the South, reformers believed "the stunting effects of child labor among whites promoted 'race suicide.'"[8] In the North, comparable fears were fomented by the presence of immigrant youth and the degradation of native-born youth. Child labor was a breeding ground for disaffection, and it was hardly the proper classroom for the inculcation of Puritan values. The enormous influx of immigrants into New England, the declining birth rates among native-born residents, and the apparent consequences of exploitation raised by disruptive strikes and attendant radicalism all encouraged reform. But regional pride was also at work. Curtis Guild, the former governor of Massachusetts and a child labor advocate, expressed it forcefully: he wanted Massachusetts "not merely to lead in some things, but to lead in all things that make for the protection of the child, for the protection of the home."[9]

Was child labor under control in the North, as Rabbi Wise had contended? In May and June 1909, Lewis Hine traveled among the mill villages of New England to find out what northern conditions were and how they compared with those in the South. Southern mills faced few or no regulatory limitations on the use of child labor. In the North, by contrast, regulations had been in place for a hundred years. Were they enforced? What was their effect? With the extensive history of regulating the age of working children, did other concerns arise—such as hours of labor, or types of industry, or quality of documentary proof of age requirements—to focus concerns about child labor. Would Hine find opportunities to photograph northern standards violated by northern labor practices? Would he be able to make images of northern children in a comparable state of misery as those he created in the South?

## RHODE ISLAND MILL VILLAGES

Hine began his New England survey with brief stops in the mill villages of Rhode Island: Anthony, Arkwright, Fiskeville, Natick, Phoenix, and River Point. The Pawtuxet

Valley was a good starting point in the search for working children: there were probably more spindles in operation within a fifty-mile radius of Providence than anywhere
else in the world, and Rhode Island had the region's highest percentage of child laborers working in cotton textile mills. Additionally, Rhode Island was one of the most
ethnically diverse New England states and had the highest proportion of urban dwellers in the nation.

Hine's photographs and captions highlighted the ethnic diversity among Rhode
Island's child laborers. In the Quidwick Company Mill in the village of Anthony, Hine
created an image of a young Polish spinner, Willie, taking his "noon rest in a doffer-
box" (fig. 3.1). Willie had literally put himself into the work, leaving concerned viewers
to wonder whether doffing spindles was a sound way to introduce foreign-born children to their adopted nation and whether mill working conditions uplifted or debili-
tated "the race" and the future of American civilization.

Hine photographed children not only in the workplace but also in the neighborhoods that surrounded the mills and factories, contextualizing and drawing together
various strains of social anxiety. Rhode Island and southeastern Massachusetts had
high concentrations of Portuguese and Cape Verdean immigrants. In River Point,
Rhode Island, Hine photographed "two Portuguese girls" who worked in the Royal
Mill; the simple image shows the young women standing against a clapboard house
with cement foundation (fig. 3.2). Hine's caption reports that the younger of the two,
Mary Fartado, "works on lace" and that both girls had been working in the mill for
three years. It seems little was done during that time to begin the process of their
Americanization: Hine states that "they do not speak English." They probably had no
need to, insulated as they were within their own ethnic settlement in River Point.

As he had in his photographs of southern mill workers the previous year, Hine often
positioned his Rhode Island subjects so that they stood next to, and were dwarfed by,
the machines they worked on; he used this juxtaposition to amplify the message that
"this child does not belong here." If he found a promising subject, Hine took more
than one photograph, working with his subject to create the desired impression. At
the Interlaken Mill in Arkwright, Rhode Island, Hine twice photographed "a beautiful
young spinner and doffer" who "looked 12 yrs. old" and had worked in the mill for
one year. In both images, Hine has positioned the girl beside a spinning frame, high-
lighting the disparity between the child, with her "hectic flush caused by warm, close
atmosphere," and the gigantic frame. In one photograph Hine uses a medium-range
focus to show more of the girl, who is standing with her hand on the machine.[10] The
image is informative but not as powerful as the second photograph (fig. 3.3), in which
he pulls the girl into closer focus and sets the machine at a slightly different angle so
that it seems to go on forever into the distance, its repetitive spindles blurring as it
disappears behind the central figure of the young woman.

FIGURE 3.1. "One of the young spinners in the Quidwick Co. Mill." (Hine 0669)

FIGURE 3.2. "Two Portuguese girls working in Royal Mill." (Hine 0689)

FIGURE 3.3. "A beautiful young spinner and doffer in Interlaken Mill." (Hine 0675)

Most mills and their agents were not forthcoming with information about child laborers, so Hine often had to be creative in his efforts to find photographic subjects. Failing to gain access to the inside of Westerly Mills, Hine was forced to photograph only those workers who were entering or leaving the mills. Hine tried another approach as well. At the Lorraine Manufacturing Company in Westerly, Hine approached William A. Wilcox, a "broker who issues working certificates there," and asked if he could secure a certificate to work in the cotton mill for the son of a friend. Hine told Wilcox that the boy was "about" twelve years old, to which Wilcox replied, "There is a way, if he has reached a certain grade in school," but said no more. In his final report to the National Child Labor Committee, Hine notes that although state law contained no such exemptions, hiring underage workers was "no doubt . . . an evasion practiced in many places."[11]

When Hine journeyed to Warren, Rhode Island, he discovered "many young children employed" there and contemplated whether this was because "families move over here from Massachusetts to get the benefit of more laxity." The possibility was worthy of further investigation, Hine concluded. As he did at Westerly Mills, Hine stood outside the gates of the Warren Manufacturing Company at 6:00 a.m. and also during the evening shift change, at which times Hine "talked with, saw and photographed about a

FIGURE 3.4. "6 A.M., June 10, 1909. Boys going to work in Warren Mfg. Co." (Hine 0814)

dozen youngsters surely under 14 years of age," the legal working age in Massachusetts (fig. 3.4). For unclear reasons, Hine was unable to get photographs of the smallest ones. A more thorough investigation, he believed, "would probably reveal more" violations of state law.[12]

Hine did more than take pictures. After obtaining "the foregoing evidence, photographic, as well as visual," at the Parker Manufacturing Company in Warren, Hine set out to interview the Rhode Island state factory inspector, J. Ellery Hudson. Hine found no advocate for children's welfare in Hudson, who seemed to "be more concerned with the question of protecting himself and the employers than he is with the rights of the children." Although Hine thought that Hudson was "in sympathy with . . . modern theories about the employment of children," Hudson's remarks belied such sympathies. According to Hine, Hudson claimed that "Child Labor Conditions are better in Rhode Island than they are in any other state in the Union. The mills are here. We invite inspection and criticism, but you can't find any violations to speak of. Of course, you might run across one or two, but we prosecute them. The manufacturers have come to know we mean business." Hudson contended, however, to Hine's obvious surprise, that prosecution did not constitute a significant proportion of his duties as factory inspector. Hine incredulously notes that Hudson instead chose to "rely upon the integrity of the manufacturers for the accuracy" of his reports.[13]

Hudson defended local manufacturers, shifting responsibility to other parties. In his opinion, Rhode Island manufacturers were "not greedy," nor were they "selfish monsters any more than any other bodies of persons." According to Hudson, "75% of the responsibility for child labor rests upon the parents." The remaining responsibility rested with school authorities, because "before a child can work, he must first become truant." Furthermore, Hudson asserted his belief that if a child was "out of school it is better for them to work than to loaf around." Rhode Island manufacturers were cutting down on juvenile delinquency when they put children to work.

According to Hudson, not only were manufacturers not responsible for child labor, but "factories are not responsible for harm to children who work." In fact, the "mill is a paradise compared with the homes of 90% of the children who work." Pointing to his own situation, Hudson related that he "began in a cotton mill when I was 10 years old,—from 5:30 a.m. to 8 p.m. and was never the worse for it." If Hine needed further evidence, Hudson offered to show him the picture of Hudson's "family of eleven children," asserting again that "work doesn't hurt children." And if that was not enough, Hudson pulled out a newspaper and proudly quoted from a piece stating that "Hudson is the man who meets criticism with a smile." Hudson charged that the National Child Labor Committee was made up "of theorists," but when asked about the "facts and pictures used by the Committee," he was "non-committal" respecting their accuracy. Hine concludes in his report, "I think he is a hypocrite."[14] Hine was conscientious about the integrity of his photographs and field reports, so he was particularly sensitive to someone dismissing "facts and pictures."

## MAINE MILL TOWNS

Leaving Rhode Island, Hine took a train to Maine for a quick investigation into conditions in Sanford and Lewiston. Hine did not stay long in either place. He didn't need to. Hine took approximately a dozen photographs in Maine and created a few images that rival their southern counterparts in displaying the social cost of child labor, underage and uneducated. Many of the textile mills in New England were located in cities, so Hine often had an urban context to augment his images of children at work. In Sanford, for instance, Hine photographed a group of workers from the Goodall Worsted Company hanging out in the street at the noon hour (fig. 3.5). There is no authority in the chaos of the street, and the young boys intermingle with the men. Hine's caption notes that there were "more young boys and young girls" than appear in his photograph, but he "couldn't get the latter to pose." The caption doesn't suggest why, but Hine's image does: it would have been inappropriate for the girls to mingle

FIGURE 3.5. "Noon hour. Goodall Worsted Co." (Hine 0703)

freely among the male workers, with the youngest boy among them, front and center, smoking a cigarette. (Hine never photographed young girls smoking, though it seems certain he would have used such a photo if he had.) For girls to jostle among the men would no doubt have reflected a dangerous immodesty of casual mixing of the sexes.

Hine made direct comparisons between conditions in northern factories and those in southern mills, using southern mills as the standard by which child labor conditions were measured. For instance, he found numerous instances of underage workers while photographing in Lewiston, Maine, and in his caption to a photograph of a "few of the young workers" at the Bates Manufacturing Company in Lewiston (fig. 3.6), he notes that there were "many more and younger" workers and that the mill was "as bad as the average South Carolina mill in regard to child labor." The image itself, however, does not convey much. Hine framed it in typical fashion, juxtaposing the children in front with older workers in the background to underscore the youthfulness of the child workers. The children are laughing and enjoying one another's company. From this picture, it is difficult to extrapolate the severity of southern conditions. Hine's caption asks viewers to take an imaginative leap that situates this mill with respect to its southern counterparts and conjures thoughts of oppression and exploitation. For

FIGURE 3.6. "6 P.M. April 23, 1909. A few of the young workers in Bates Mfg. Co." (Hine 0718)

New Englanders who were prideful of their reputation, the caption would be discomfiting, revealing the hypocrisy of pointing a finger at the South while ignoring issues in their own region or the industry that connected the two regions. Linking conditions in northern and southern mills was important, since anti–child labor advocates aimed at securing national legislation to supersede the hodgepodge of protections individual states afforded the nation's children.

One of Hine's Maine photographs and its caption create an especially disturbing portrait of child labor in New England (Fig. 3.7). Hine photographed a "small boy" named Jo, a sweeper at Hill Manufacturing Company in Lewiston, "going to work at 5:30 A.M." Jo "lives 2 miles out in the country, and his father drives him out and back every day." It is unclear whether the man standing beside Jo is his father. It is less likely that the man behind them is a relative. Jo had been working at the factory for two years, according to Hine, but he looks very young in the photo. The caption makes us wonder whether the father was living off the boy's earnings, meager though they would have been. Hine further disconcerts the viewer and evokes empathy by telling us that Jo "said he didn't know how old he was and he couldn't spell his name." Clearly, child labor was not a good way to nurture the rising generation.

FIGURE 3.7. "Going to work at 5:30 A.M." (Hine 0709)

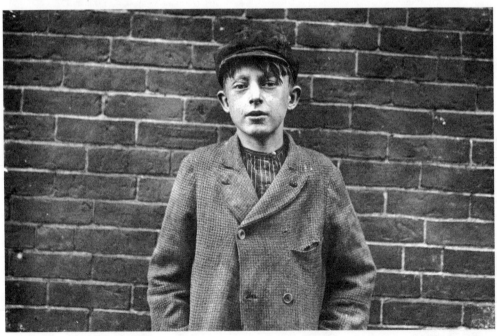

FIGURE 3.8. "One of the boys working in Bates Mfg. Co." (Hine 0714)

Hine replayed these same concerns throughout his next few images from Maine. He photographed a boy who worked at the Bates Manufacturing Company in Lewiston (fig. 3.8) and in the accompanying caption states that the boy had "been there 3 years." Again the viewer is forced to confront a child worker's young age. The boy's clothes are tattered and worn, missing buttons, and draped awkwardly on his body. He is looking straight at the camera with sleepy eyes. The boy is representative of the poverty that was common to both the northern and southern regions. He is also foreign-born: instead of learning the language and the new culture, the boy, we are told, "cannot speak English." This image and Hine's other Maine photographs reflect New Englanders' anxieties about child labor being a barrier to the assimilation of immigrant children into American society.

## VERMONT HAMLETS

Hine continued his tour of New England, traveling into Vermont after leaving Maine. The state of Vermont was not known for its industrial manufacturing or for its ethnic diversity, but Hine found ample opportunities to carry on his project. In his caption for a photograph of boys who worked at the Chace Cotton Mill in Burlington (fig. 3.9), he highlights the number of foreign-born among his subjects, noting that "only a few could speak English," and that more boys work in the mill than are shown in the photograph. Thus he taps once again into concerns about the influx of immigrants. These children were not newcomers to factory life. Hine's caption reports that "many of the smallest ones have been there from one to three years." The boys in the photo are coming of age, with the cigarettes in their mouths confirming our perception that they are getting old before their time. Hine's impressive ability to collect their names provides detail and specificity to his story. The boys were apparently cooperative, as it would have been no small task to organize them for the shot and gather their names for the caption.

Lewis Hine occasionally made photographs that showed the dignity and strength of the children. The powerful image of three young women (fig. 3.10) taken in the drawing-in room of the American Woolen Company in Winooski is one. In a well-balanced composition Hine has posed two of the girls so that each has her left hand resting on the shoulder of the girl to her left, while the girl at far right has her left hand on her hip. His caption notes his suspicion that two of the girls were less than fourteen years old, but they were not the youngest girls he saw that day; he counted twenty girls "like these and younger" going into the factory at seven in the morning. What the caption does not tell us is that this was Hine's third version of this photograph. He tried it with various arrangements of different girls and slight modifications to their poses. Hine knew what he wanted and worked the idea until he had constructed an image to serve his cause.

FIGURE 3.9. "All these small boys, and more." (Hine 0730)

Not all children laboring in the textile industry faced dead-end jobs; some were learning valuable skills that supported a more comfortable future. There were two basic processes for spinning yarn: mule spinning and ring spinning. Mule spinning, the older of the two, was used to spin finer grades of yarn and was found extensively throughout New England. Ring spinning, the only process used in the South, was also widespread in the North.[15] Mule spinning was a career path that was often passed down from father to son. Along with loom fixing, it was one of the most skilled jobs in New England cotton mills. Ring spinners could be replaced with relative ease by children or adults, but becoming a skilled mule spinner required many years of apprenticeship. This gave mule spinners more power in the factories and made them among the best-paid workers in the cotton mills of the North. Mule spinners (always men) used children as "back-ropers" or "back boys" to help in the mule spinning process. This was how boys were introduced to the work and began their informal apprenticeship. To go from back-roper to mule spinner was no small accomplishment, and it certainly was not a sure thing. However, being a back-roper was one of the few occupations for children in textile factories that placed them on a trajectory toward becoming a skilled worker.

FIGURE 3.10. "Noon hour in drawing-in room of American Woolen Co." (Hine 0728)

FIGURE 3.11. "Jo Bodeon. A 'back-roper' in mule room." (Hine 0742)

In Burlington, Vermont, Hine made multiple photographs of back boys working in the mule room of the Chace Cotton Mill. He took long-range, medium-focus, and close-up shots of the boys alongside their spinning mules, which were sometimes capable of running a thousand spindles at once. Hine made two frontal portraits of a back-roper named Jo Bodeon; both images are balanced by light from the nearby windows on one side of Jo, who stands in central focus, and white roving on the spinning frame on the other side. In the first image, Hine photographed Jo in a medium-range shot (fig. 3.11). The image is very crisp, with a sharp focus throughout the composition. The perspective is essentially comparative and emphasizes the large size of the machine and the small size of Jo.

The second photograph offers a tighter perspective on Jo (fig. 3.12). He is also in sharp focus here, but unlike in figure 3.11, the windows on the left and the spinning frame on the right are out of focus. Hine allows us to look more closely at Jo, and as we do we lose the perception that he is dwarfed within a cavernous factory. The power of this image is in the detail, in Jo's countenance and his vacant gaze. He stands alone, his hair slightly mussed, his face without a smile, numb, dazed, scared. There is the worn shirt, with each button a different color and a safety pin holding the center. The suspender string holds his pants up, and the cotton fuzz that seems to radiate from his worn wool pants projects privation. The details that Hine captures in this photograph make it a vital image for reform.

Not only did mule spinning require a much higher skill level than ring spinning, but it also produced a much finer thread and one of much higher quality. Manufacturers had tried to automate the mule at least since the 1840s, but its complexity as a machine made automation difficult. The low end of the spinning industry migrated south after the Civil War and depended solely on ring spinning, which was technologically very conducive to the employment of children. Most spinners in New England worked as ring spinners, but the region retained the more specialized segments of the industry, including the mule spinners. The small number of children who worked as back-ropers helping the mule spinners learned from the most skilled workers in the industry.

## NEW HAMPSHIRE CITIES AND TOWNS

In May 1909, Hine traveled to Dover, New Hampshire, to investigate conditions at the Cocheco Manufacturing Company. Because of the size of the Cocheco plant, which had employees "scattered" among five large mills, Hine at first had some difficulty locating child labor violations, but he persisted. "By constant watching at the several gates, morning, noon and night for several days," Hine identified more than two dozen child workers, "some of them being apparently 10 and 11 years old." Hine found

FIGURE 3.12. "Name: Jo Bodeon. A 'back-roper.'" (Hine 0743)

conditions that looked like those in some of the southern mills, and he needed to photograph those conditions.

Unable to enter the mill complex, Hine took candid photographs of children coming from and going to work at the Cocheco plant. When he made a photograph of young girls going to work in Mill No. 1 (fig. 3.13), he did not line up or pose his subjects, nor did he gather their names. It is not clear whether the girls even saw Hine. The photo shows two clusters of girls—a pair of girls walk in the direction of the camera, passing by three girls who are standing and talking at the corner of a building. The building corner, the archway, the windows, and the cobblestone create an unusual perspective. None of the young women are wearing clothes that indicate their occupations; if not for Hine's caption, they might be mistaken for schoolgirls. The man whom we see in silhouette in the darkened doorway at left lends an ominous air to the scene.

Hine went beyond simply identifying children at work; sometimes he followed them home to speak with parents or to test the veracity of the children's stories. Hine looked for children on the streets or at the "ball grounds" during the evening hours

FIGURE 3.13. "Young girls going to work in Mill No. 1 Cocheco Mfg. Co." (Hine 0775-A)

and on holidays. In these city sojourns he spoke with many children, but most were "alert enough to put themselves at 14 when questioned"—although, Hine asserts, "I am positive that many were not." On this occasion, Hine reports, he "traced" some of the children to their homes. Language barriers and parental reticence inhibited his investigation, and "the only mother who could understand English," he notes, "said her boy was 14—but I know better."[16]

Hine was right to think he knew better, as the communities in which he photographed did not always share his reform sensibilities. Parents and children collaborated with priests and others in deceiving local authorities so that underage children could work. Many children went into the factories as "'helpers' to mothers, fathers, brothers and sisters," according to Hine, "but they work regularly."[17] Sometimes at noon a small army of young children could be seen delivering supper to their parents, siblings, or other relatives in the mill. They began as helpers but soon found their labor called upon for a host of activities, from sweeping floors to doffing spinning frames.

On one occasion, Hine photographed Charles Chasse, a small boy who had worked for several months as a back-roper at the Great Falls Manufacturing Company in Somersworth, New Hampshire (fig. 3.14). Hine later visited the boy's home; when he

FIGURE 3.14. "Small boy works as helper—backroping in Great Falls Mfg. Co." (Hine 0767)

inquired about Charles, "his mother very reluctantly admitted he was working, but said he is past 14 years of age." When Hine met the boy again on the street, Charles confessed that "he is to be 14 next month, when he hopes to get his 'school papers.'" Parents often coached their children to say that they were of legal working age, although children couldn't always be counted on to follow through on the deception.

Great Falls Manufacturing Company was a large concern, with four mills and four gates that Hine watched for evidence of child labor. Hine observed "a number of children under 14 but couldn't estimate how many." Some of the children appeared to be only eleven or twelve years old. When queried by Hine, a number of them said that they merely "helped," with Hine noting that they did so "regularly and in school hours." Once again, Hine took down some of their names and addresses and subsequently visited their homes. These visits, however, offered him "little satisfaction," as most of the parents, "like the children," were "'wise' and [said their child was] '14 years old.'"[18]

In Penacook, New Hampshire, a small town with two cotton mills, Hine "expected to find many violations" but was "very agreeably disappointed." Hine gained access to both the Boscawen Manufacturing Company and the New Hampshire Spinning Mills. At the Boscawen Mills, which had seventy employees, the superintendent let Hine "go through the whole mill, alone," but Hine found no employee under eighteen years of age. At the New Hampshire Spinning Mills, where Hine was also allowed total access to the mill property, he uncovered only "three doffer boys . . . on the border line of 14" among the 150 employees and was unsure whether these were violations or not. On the whole, however, conditions at the Penacook mills were "very good."[19] Hine did not take any photographs during these mill visits. In most instances, favorable working conditions did not make for successful child labor images.

When Hine visited Suncook, New Hampshire, he was pleasantly surprised by the local mill conditions—they were the best he had seen in New England. The three mills in Suncook had recently consolidated under the management of the Suncook Mills. As in Penacook, Hine enjoyed "exceptional opportunities to see the actual working conditions in all three mills." Moving through part of the mills alone, and then accompanied by management, Hine found "only three boys and 1 girl that I was reasonably sure were under 14." This, Hine notes, was a "very good record for a large mill in this state." Employing twelve hundred "hands," Suncook was a sizable operation with "unusual conditions for a large N.H. Mill." The only photograph Hine made at Suncook was of a group of boys standing in front of the mill (fig. 3.15); his caption notes that "sanitary conditions here are exceptionally good, for this section." In his report on the New Hampshire investigation, Hine states that "the living and working conditions of the people is being looked after [at Suncook] better than in any place I have seen in New England." He commends the efforts of the recently deceased superintendent, Mr. Whitten, and notes that Whitten's son was committed to carrying on his father's work. Hine recommends further investigation of conditions at the Suncook Mills, "to see something that is being done on the positive side of the industrial problem."[20] This was a departure from Hine's usual practice of showing what was wrong in the hope that it would be made right.

The situation at the giant Amoskeag Manufacturing Company in Manchester was markedly different from what Hine witnessed in Penacook and Suncook. Visiting Manchester between May 21 and May 26, 1909, Hine quickly discovered that Amoskeag did not allow visitors into its factories, and photographers "were relegated to the outside." Using some form of subterfuge, Hine nevertheless went in "with the 'hands' at 6:00 a.m. and at noon on various days of that week." Once inside, he gained access to the principal spinning rooms, where he found a number of children at work or beginning work. On his first day inside the factory, Hine photographed a group of boys and girls—though "not the smallest ones"—that he found working in the spinning room (fig. 3.16). In his caption Hine states that he took the photo in a "hallway"; the photograph itself provides few clues as to its location, as the image is tightly framed, and the children are surrounded by darkness. In conversing with the young workers, he was able to get their names and addresses, which he included in the caption. One

FIGURE 3.15. "Group of boys who work in Suncock [*sic*] (N.H.) Mills." (Hine 0753)

interesting component of this photograph is the stance adopted by three of the boys in front—each with his arms crossed, each taking a bit of pride in his work.

Hine was somewhat freer in his efforts outside the gates of the Amoskeag mills, where he posed "in the guise of a postal card man" taking photographs (fig. 3.17). The children he photographed outside the factory gates were in "many cases" the children Hine had seen working inside, or they were children who "told me they were working" in the mills.[21]

Hine clearly worried about his credibility as a social documentary photographer, fearing that people would challenge the veracity of the images he created. In his field report, he asserts that "these photos are, therefore, not to be refuted by those who would say that the children that come out of the mills are the 'dinner-carriers.'" As further verification of his work, Hine reports that E. W. Lord, secretary of the New England chapter of the National Child Labor Committee, "was with me several times[,] and one evening" Hine counted thirty boys and girls that he considered under fourteen years of age coming out of one of the Amoskeag gates. Moreover, Hine notes, "there are many other gates." Posted at one of the gates, Hine counted more than a dozen children going into the mills during the morning and noon hours and leaving the mills at 6:00 p.m. Regarding the appearance of "dinner-carriers," although Hine tried to avoid taking their photographs, he nevertheless believed it "a mistake to permit any children

FIGURE 3.16. "A few of the small girls and boys (not the smallest ones)." (Hine 0748)

FIGURE 3.17. "Noon, Saturday, May 22, 1909. Young girls working in Amoskeag mills." (Hine 0756)

FIGURE 3.18. "Noon hour, May 26, 1909. These boys and many smaller ones work in Amoskeag Mfg. Co." (Hine 0792)

to go inside the gates even to carry dinners," for the mere "presence of a child under 14 in the mill yard or mill should be prima facie evidence of employment." Many children began their careers in the textile industry as dinner carriers.

Hine produced a dramatic series of photographs of Amoskeag workers walking home at day's end. Image after image documents this daily procession of laborers returning to their homes in the West Side of Manchester, where the city's large French Canadian community lived. In many respects, the photographs show this community on the march, with children very much a part of the proceedings. Sometimes younger children came out to watch the parade of labor, but for the young factory workers in Manchester, partaking in this daily ritual with their peers and parents, their neighbors and friends, affirmed their inclusion in this community (fig. 3.18). French Canadian families throughout New England valued their children's contribution to family income more than they did their children's education.

The strongest influence on children's educational advancement was "the attitude of the parent toward the school." Parents' attitudes, in turn, were heavily influenced by their own ethnic social and cultural values. Three-quarters of English, Irish, Scottish, German, and various other non-English-speaking fathers in Manchester sought to keep their children in school, while two-thirds of Canadian fathers had "little or no faith in the value of school training." That more than half of Manchester's immigrant population was from Canada made this particular cultural value a central part of the city's "educational problem."[22] The confidence placed in factory labor over education suggests the widespread belief that industry fit into the population's future, as it had in the past. It was a known and well-worn path. That the industry was relocating geographically did not seem to figure heavily in families' calculations. Children expected that they would labor as their parents had labored before them, and that the industry would go on as it had in the past, providing employment for generations of workers.

Few of the children who left school for factory life had had much educational success, or many expectations of success. Child labor investigators for the Women's Educational and Industrial Union (WEIU) asked working children in Manchester why they had left school,[23] and 90 percent of their answers fell into one of three responses—"disliked school," "economic need," and "wanted to work, or to earn." According to the WEIU's report, more than a third of the children replied that they simply did not like school, leading some authorities to consider this a clear "indication that the school is not adapted to meet the needs of this type of pupil." The desire to work or earn expressed by one-fourth of the children, they noted, should be a central part of considerations in "planning the new school program." Almost 30 percent of the children claimed that they had left school out of economic need. Investigators refrained from endorsing an easy solution such as mothers' pensions.[24] Instead, they sought to validate the children's claims of financial need by further investigating the economic conditions of the families.[25] After comparing family size, total income, and number of wage earners in families, investigators found ample evidence to "justify claims that necessity prompted the early entry into industry."[26] But the children's entry into the workforce ensured at the same time that their expectations for success would remain low.

Investigators found a very high level of turnover among the child laborers in their study of children's employment patterns. When analyzing the impact of wages on the decisions child laborers made, investigators discovered a high level of what they called "shifting," or moving from one job to another in pursuit of wage gains. There were two ways in which a child could obtain better wages. First, the child found a good job and "stuck with it." Second, the child changed positions to secure a higher wage. Investigators called the latter strategy into question. In some instances, the initial wage was higher in the new job, but ultimately the gain was less than that earned by children who stuck with their first job. In many cases, however, such as that of a boy who increased his weekly wages from $6.50 to $12.50, shifting resulted in real economic success for the child. So shifting worked out sometimes and didn't at others; but children, like their parents, still sought to secure what advantages they could through their freedom to vote with their feet, so to speak, and move to another job.[27]

FIGURE 3.19. "Boy with bare arms, Fred Normandin." (Hine 0788)

   "Shifting" was not always the choice of the child laborer. There were instances of children being dismissed for "fooling" around or simply written up as "getting fired." Fred Normandin was one of those children whose work habits were less than stellar (fig. 3.19). According to the Amoskeag Manufacturing Company's employee records, Fred was hired in October 1911 to work as a bobbin boy. One week later, Fred was fired because he had "stayed out." We don't know why Fred stayed out—perhaps he disliked the work and was shifting to new employment, or perhaps he had difficulty showing up to work for some reason. Some employees, having entered and left Amoskeag employment numerous times over the years, had dozens of employee cards on file. The cards asked whether the company should ever consider hiring the individual again. In Fred's case, the answer was an emphatic "No." Fortunately, there were the shoe factories.[28]

   Hine's image of George Crossley (fig. 3.20) is the only one in the textile group of New England photographs in which he suggests that child textile workers are performing night work. Hine photographed George as he was going into work at 6:00 p.m. According to Hine, George had been working an all-night shift for the previous five months. Still in his knickers, George labored through the night. George was one of the individuals whose persistence in the same place of employment paid off—if not

FIGURE 3.20. "Going to work on the night shift." (Hine 0796)

in wages, then certainly in his future career. George's employee record shows that he continued to labor for Amoskeag after this image was created and in time was given a new opportunity: in July 1911 he was hired to work as a back boy in the mule spinning room in Mill No. 5 of the Southern Division. George's employee record also has a copy of his birth certificate, which by 1911 was required under the new law strengthening documentation requirements for getting permission to work.

New England prided itself on its traditions of reform, so it came as a shock to some when Hine's photographs provided evidence that child labor was very much present in the region. Every now and again during his work in New England, Hine created images (such as the photograph of Addie Card discussed in chapter 1) that rivaled those he made in the South. Similar to Hine's photograph of a little spinner girl in an Augusta, Georgia, mill is his image of an anonymous young girl who appears to be "11 or 12 yrs. old" and is standing by a spinning frame in the Amoskeag mill (fig. 3.21). This is one of the genres Hine commonly used: the child next to the machine. The white roving on the spinning frame directs the viewer's eye to the girl at center, while the machine provides a means to gauge the girl's height, which Hine's caption gives as 48 inches. (Needing to evaluate and record his findings rapidly, Hine used the buttons on his

FIGURE 3.21. "Name: Little girl (48 inches high) work in Amoskeag Mfg. Co." (Hine 0786)

shirt to provide an accurate measurement of a child's height.) The girl is barely tall enough to see over the spinning frame rollers. Hine uses this structure to juxtapose innocence with industry.

The image appears straightforward enough, yet there are many mysteries. Was the girl 11 years old, or was she 12? Did she have working papers? Was she attending school? Did she work at this machine? If not, where did she work, and in what capacity? Textually, the most puzzling element in this image is the man standing in the background, looking down on the girl. Who is he? Is he an overseer? Another "hand"? A loom fixer? A parent? How do we explain this man, whose presence creates a tension that gives the image its power? While the little girl is picking at the thread being spun, or is posed to look as though she is, the man is not working; he is watching, with hands in pockets, intentionally or unintentionally suggesting that someone benefits from the labor of this child—that because she works, someone else doesn't. By tying the man to the girl, Hine makes him complicit in her labor. If the man was an overseer of some sort, his complicity spoke ill of the Amoskeag Company, which failed to control what was probably illegal activity. His mustache and cap also give the impression that he is ethnic, a "foreigner." His ethnic profile would have contributed to the belief that immigrant newcomers, cut off as they were in worlds of their own, didn't understand American standards and values, or conditions that would need to be rectified before citizenship was bestowed.

Even with the help of E. W. Lord, Hine found it "very difficult to estimate the number of violations in such a large corporation," but he believed that the photos showed "*many* violations [Hine's emphasis]." Unable to quantify the precise extent of violations in Manchester, Hine nevertheless expresses his indignation but he didn't rely solely on the photographs to speak. "With the photographs before us," Hine writes, "it is interesting to note that the Amoskeag Corporation is one of those New England mills referred to by Senator Gore recently and quoted as paying dividends from 30 to 60%. In answer to the charges, the Amoskeag people said their profits came largely from *real estate* [Hine's emphasis]." "Conditions in all these states, except Massachusetts," he concludes of his sweep through New England, "are much below par." There were enough violations in New England mills to give one pause when it came time to celebrate the advances made in the region.

The National Child Labor Committee sought federal legislation to equalize legal protections for children from one state to another. For Hine, the question continued to be, "Why this double standard?" In a poster (fig. 3.22) Hine produced in 1913, he set up a direct comparison between the New England and southern regions. Noting that "one New England corporation" owned cotton mills in both Massachusetts and Georgia, Hine compared the regulations that children labored under in the different states. In the Massachusetts mill, immigrant children sixteen and upward worked

FIGURE 3.22. "[Exhibit panel] Location: Dallas, Texas." (Hine 3621)

ten-hour days. In the Georgia mill, by contrast, native-born children ten and over labored eleven hours per day. The images that Hine collected in his first study of New England textile workers showed violations of child labor regulations. His photographs established that child labor was present in both the North and the South, although clearly the conditions differed between the regions: the South had competitive advantages, with limited state intervention into matters of child labor or compulsory schooling, and would not surrender those advantages easily.

Hine recommended that "intensive work should be put upon every one of [the New England states] at the earliest opportunity."[29] In 1906 Congress ordered the

Department of Commerce and Labor to undertake an investigation into the conditions of woman and child workers across the United States, which ultimately produced a massive nineteen-volume report.[30] The report was awaited with great anticipation. Many people expected it to provide evidence that "the worst of woman and child labor are found in the cotton mills in the South," but the report provided page after page of evidence that if women and children's working conditions were worse in the South than anywhere else, the difference was only one of degree. Commissioner of Labor Charles P. Neill believed that the report would "revive this agitation and show, that not alone in the interest of humanity, but to preserve the vitality of the race, there must be a prohibition against the working of women beyond certain stipulated hours and a minimum age limit must be fixed for the employment of children in factories."[31]

In a letter written to a former associate at the Ethical Culture School after his New England trip, Hine stated, "My child labor photos have already set the authorities to work to see 'if such things can be possible.' They try to get around them by crying 'fake,' but therein lies the value of date & a witness. My 'sociological horizon' broadens hourly."[32] In his first sweep through the region, Hine took the skills he had learned in the Pittsburgh Survey and put them to work in documenting a generation of children throughout New England who labored daily in the textile industry they helped create.

## CHAPTER 4

# Exhibiting Child Labor

———

Lewis W. Hine did not exhibit his work in traditional museums at this stage in his career, but he did display his photographs to the public. Beginning with the Pittsburgh Survey, Hine's visual contributions to social surveys were essential components of such studies, substantiating statistical evidence and validating narration.[1] Hine's photographs allowed viewers to authenticate "with their own two eyes" the evidence placed before them. Visual evidence grew in use and importance for all types of reform activity. Reform initiatives that hoped to get the public's attention came to depend increasingly on visual evidence, analysis, and communication. The growing value of the visual in social reform was part of a larger cultural preference for pictures, images, illustrations, movies, charts, graphs, cartoons, and other graphic forms, and it reflected the emergence of a more literate visual society.[2]

Progressive reformers embraced the exhibit as a principal means of communicating to a broad public and a useful tool in creating a community among professional reformers themselves. Reformers from local and national organizations and from both public and private associations relied on the power of graphic imagery to document their causes and advertise their arguments for change. Lewis W. Hine's photographs were ideally suited to the visual needs of Progressive reform organizations, and they fit in perfectly at Harvard professor Francis Peabody's new Social Museum, a museum of reform that collected and exhibited photographs, illustrations, and other visual and textual materials documenting social conditions in the United States and elsewhere.[3] The Pittsburgh Survey showed how to use visual evidence; the exhibition became a tool to display such evidence, and the social museum emerged as an institution for circulating, reusing, studying, and teaching the evidence.

## CHAPTER FOUR

### EXHIBITING FOR SOCIAL REFORM

By the turn of the century, many people saw the photograph as an essential tool for mass communication. Photography had already shown its utility in the world of exhibitions. At the 1893 Chicago World's Fair, organizers used photography extensively and in novel forms to communicate to the massive audiences. The historian Julie K. Brown, in *Contesting Images: Photography and the World's Columbian Exposition*, has argued that photography was "essential to the staging" of the exposition, "providing both the necessary images for display and a means for making images of the experience of the event itself." Many people put great faith in photography as a "tool for science, an agent of social change, and a replica of things not present."[4] The photograph could reach a polyglot population like no other form of communication.

The successful use of photographs in important exhibitions affirmed photography's vital role in modern visual culture. The historian Shawn Michelle Smith, in *Photography on the Color Line: W. E. B. Du Bois, Race, and Visual Culture,* traced the visual meaning of the color line by examining Du Bois's Exhibit of American Negroes at the Paris World's Fair in 1900. In this prizewinning exhibit, Du Bois first introduced the idea that the problem of the color line was *the* issue of the twentieth century.[5] Lawrence Veiller, a social reformer, housing expert, and volunteer for the Charity Organization Society, created the Tenement-House Exhibition of 1899 to raise awareness of the terrible housing problem in New York City. For two weeks, visitors to an old Fifth Avenue building experienced an overwhelming display of models, illustrations, factual panels, dozens of maps, and more than a thousand photographs that was intended to arouse the public into taking action against tenements. The Danish American social reformer and documentary photographer Jacob Riis recalled that "rich and poor came to see that speaking record of a city's sorry plight." The graphic evidence was undeniable, and "not to understand after one look . . . was to declare oneself a dullard." The exhibit had far-reaching influence and helped secure passage of the New York State Tenement House Law of 1901, which became a model for housing legislation in cities across the nation.[6] The Tenement-House Exhibition also became a model for the use of exhibits to enlighten the public on critical social issues.

Progressive reformers brought together the perceived objectivity of the photograph and the "scientific social survey" to chart a new course for reform. Reformers intended that the information and data gained though social surveys be recorded, analyzed, and then shared. Russell Sage Foundation director Shelby Millard Harrison, in his study of the development and spread of social surveys for the foundation, noted that the research methodologies embodied in the social survey "took definite form" following the Pittsburgh Survey. Social investigators believed that "a new type of endeavor was born which not only articulated developing needs and developing scientific tools, but also gave illustration and impetus to an idea that was destined to spread widely." Jacob Riis's study *How the Other Half Lives* (1890) had already shown the power of photographs. Reformers were treated to the birth of urban sociology in *Hull-House Maps and Papers* (1895), a book by the residents of Hull House, and *The Philadelphia Negro* by W. E. B. Du Bois (1899); those works showed how valuable in-depth neighborhood surveys were in fully understanding an area. The Pittsburgh Survey conjoined the photographic and social survey approaches. That orientation subsequently solidified and became nearly universal in the field.[7]

Progressive reformers embraced the social survey as a means to obtaining a better understanding of social situations and problems and the exhibition as a way of communicating that information to a broad public. The survey was very much a cooperative endeavor to apply scientific methods to the study and treatment of social problems within defined geographical boundaries. Social researchers formulated the questions for study, investigated and analyzed "pertinent facts," and drew together "warranted generalizations." Specialists from various fields, physicians, city planners, and social workers helped in "bringing problems down to human terms and in prescribing or planning treatment." Finally, vital to the success of these efforts was the work of "the journalist and publicity worker in interpreting facts and new knowledge in terms of human experience and presenting them in ways which will engage the attention and stimulate democratic action." Public exhibits became important vehicles for developing "the common knowledge of the community," spreading facts, drawing conclusions, and making recommendations for intelligent coordinated action.[8]

The problems on display in public exhibits often reflected the social divide between the observer and the observed. These exhibitions offered the emerging professional class an opportunity to look at the foreign-born and observe working-class conditions and patterns of behavior. In description, categorization, and prescription, reform-minded experts distinguished themselves from those on display. They saw themselves as nonpartisan, simply applying their particular specialization to diagnose and treat the cities and neighborhoods and their inhabitants. However, even as they set themselves apart, these same experts were often advancing the class interests of their subjects through their sincere calls for improvement of working or living conditions, access to meaningful education, and better health. Hine, like his colleagues, wanted to improve social conditions and committed to playing his part by creating graphic images to frame the discourse on child labor and advance the cause of child welfare (fig. 4.1). Hine's exhibits delineated the boundaries structuring a modern middle-class understanding of childhood and child labor, which was juxtaposed with working-class practices, values, and experiences.

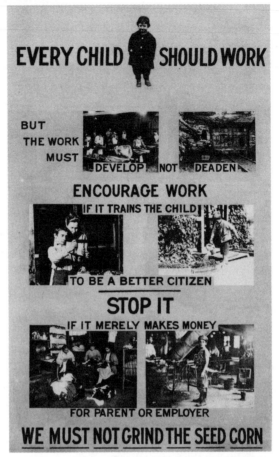

FIGURE 4.1. "Exhibit panel." (Hine 3782)

## LEWIS HINE AND THE BOSTON-1915 MOVEMENT

In 1909 Hine photographed for an exhibition in Boston whose focus went far beyond issues of child labor reform. The "1915" Boston Exposition approached social issues as complicated and multidimensional problems that would require and benefit from the contributions of people with expertise in various aspects of urban life, such as housing, education, recreation, health, and law. Hine produced images that identified and defined urban social problems, while also creating images that aimed at showing the positive steps that could be taken to address those problems.

In Boston, Hine worked with the Boston-1915 movement, a broad coalition that attempted to treat social problems more efficiently and holistically. The stated aim of the Boston-1915 movement was to create a better Boston. Rather than leaving the fulfillment of that aim to some indefinite time in the future, the movement set 1915 as the year for its attainment. The basic idea of Boston-1915 was to "apply to the activities of the city what every well-managed business partnership applies to its factory, shop or store—to have every department working in close co-operation with every other, in order that results may be produced most quickly, economically and satisfactorily."[9]

The Boston-1915 movement set far-reaching goals. The movement's organizers challenged Bostonians to unite behind reform efforts by dedicating themselves to creating a livable city. Those who attended the first meeting of the Boston-1915 movement committed themselves to the proposition that "it shall be possible for a willing worker earning an average wage to live, himself and his family, healthfully and comfortably; to bring up his children in good surroundings; to educate them so that they may be truly useful, good citizens; and to lay aside enough to provide for himself and his wife in their old age." With "a million hands and one will," the movement strove to establish a well-planned and well-developed city; better low-rental housing; greater health protections; more playgrounds in crowded districts; more public buildings with libraries, gymnasiums, and meeting halls; better relations between workers and employers; the use of schools as social centers outside of school hours; and a school system that could provide guidance to girls and boys in finding out what work they were best at doing and then train them to do it well. Advocates saw the Boston-1915 movement as "a protest against the system that would wear out a human being as a worker, and then care for him as a pauper." A clean, beautiful, honest, and well-governed Boston would be a better place to live for all of the city's inhabitants.[10] The objectives of the Boston-1915 movement stood in contrast to the impressive White City built for the Chicago World's Fair in 1893—Boston-1915 did not seek its fruition in "transient buildings of stucco, to delight the eye for a few months." Its promoters expected six years of solid achievement by the people of Boston, so that their city might "be the finest city on earth" by the year 1915.[11] The movement was optimistic when it came to the powers of reform.

The Boston-1915 movement was conceived by Edward Albert Filene, the wealthy department store owner and Progressive reformer, and was modeled philosophically after the Greater Berlin movement, which had tried to integrate the towns and cities surrounding Berlin into a comprehensive and inclusive plan for the metropolitan region. Planning was a central focus of Boston-1915, born of a determination to rationalize urban industrial life with a six-year plan of civic improvement built upon the coordination and efficient administration of city space and city services. In 1909, urban planning emerged as a key discipline in social reform. The first National Conference on City Planning was held in Washington that year. Also in 1909, Daniel Burnham published the Chicago Plan, showing the growing sophistication of urban planning in the United States. The Boston-1915 movement was a public exercise in the intellectual and cultural refashioning of Boston into a new urban community, a metropolitan Boston.[12]

To nurture broad participation, the Boston-1915 Committee organized wide-ranging programming designed to be inclusive of diverse interests and varied for different audiences in the city. The first year's program included not only the display of a large group of exhibits on social conditions in the city but also a public lecture series; a citywide pageant; the launch of *New Boston,* a monthly journal chronicling "progress in developing a greater and finer city"; and a series of civic rallies throughout greater Boston hosted by Boston mayor John Francis "Honey Fitz" Fitzgerald. The exposition component of the plan was staged to "make the Boston-1915 movement better understood." To ensure that the movement's research and knowledge would be accessible to a broad public, it was critical that the "1915" Boston Exposition "show the people of Boston in a graphic way the work of their own and of other cities." The exposition had an additional function, one that was in some respects more important to the spirit of the plan: "to bring the organizations working in the city into closer acquaintanceship" and to foster "greater co-operation among the forces for city development"—Progressive forces with democratic aspirations, coordinating forces with a new vision of urban administration.[13] The exposition was the social medium drawing people together.

At the core of the movement were organizations and associations that had the greatest interest in child welfare, but the intent was always to draw new people into that circle of service to children. John L. Sewall, editor of *New Boston,* maintained that "improvement in conditions of child-life is at the beginning of all progress."[14] The movement attempted to coordinate the contributions of thirteen different working groups, including groups representing business, labor, cooperative associations, and city planning as well as charity organizations, health organizations, social settlements, educational institutions, and religious organizations. Boston-1915 sought to be cooperative and comprehensive—to join the material with the spiritual, commerce with philanthropy, beauty with utility. If executed correctly, it would create a congress of concern for child welfare that would bridge class divisions and erase distinctions between the original colonial settlers and "the latest steerage arrivals from southern or central Europe."[15] The movement drew members of the professional class together in an effort to uplift and Americanize the children of immigrant workers.

Mayor Fitzgerald staged a series of civic rallies throughout the Boston metropolitan district at which leading reformers addressed diverse audiences on the subject of their respective areas of expertise. Among the leading speakers were Clinton Rogers Woodruff, secretary of the National Municipal League; Henry B. F. Macfarland, a former commissioner of the District of Columbia; Owen R. Lovejoy, general secretary of the National Child Labor Committee; Lawrence Veiller, secretary of the National Housing Association; Paul U. Kellogg, managing editor of the *Survey;* George E. Johnson, superintendent of the Pittsburgh Playground Association; and Florence

Kelley, general secretary of the National Consumers League. Dozens of local academics, social workers, city planners, religious leaders, and labor, business, and government representatives held these rallies in diverse venues such as the Sanders Theatre in Cambridge; Navillus Hall in Dorchester; town halls in Everett, Lexington, Revere, and Watertown; the Francis Parkman School in Forest Hills; Bowditch School Hall in Jamaica Plain; Malden and Medford High Schools; Quincy's Alpha Hall; Andrews School in South Boston; the Walker and Pratt Company shops (for a noontime program) in Watertown; and Highland Hall in West Roxbury. The Central Labor Union organized an evening event at Faneuil Hall at which Paul Kellogg and Florence Kelley addressed organized labor. A final civic rally was held November 21 at the Tremont Temple and featured Charles F. Thwing, the president of Western Reserve University, and Rabbi Stephen S. Wise of New York, among others.[16]

Boston-1915 organized an extensive four-week lecture series with an impressive lineup of speakers. Archbishop O'Connell of Boston, the settlement house worker Jane Addams, Rabbi Stephen S. Wise, the Chicago architect Daniel Burnham, and James Duncan, vice president of the American Federation of Labor, kicked off the series on October 30. On November 5, Boston-1915 sponsored an illustrated lecture by Florence Kelley on the subject of "Worshippers and Workers at Thanksgiving and Christmas." On November 10, Everett W. Lord, secretary of the New England Child Labor Committee, presented an illustrated lecture on "Sacrificing the Children on the Altar of Industry"; that same evening, Harvard professor Francis Greenwood Peabody gave an illustrated lecture on "The Resources of Harvard's Social Museum."[17] It was an inspiring array of voices speaking out on the child labor question.

The most dramatic and democratic of the programmed events was "Cave Life to City Life," a civic pageant tracing the history of civilization from the primitive past to the urban present. The pageant was mounted at the Boston Arena on the evenings of November 10, 11, and 12; rehearsals began in early summer and involved "a thousand citizens." According to the pageant's promoters, these rehearsals had the effect of stirring community life and awakening municipal pride and a "splendid spirit of civic co-operation," thus producing what they believed was a truly "Greater Boston." Lotta A. Clark, the pageant's director, received the "invaluable assistance" of volunteers from all corners of the city, "professional men and women" who "sacrificed their own interests for the success of the pageant." Their sacrifice wasn't wholly magnanimous. In their desire to help, they exercised their professional connections in their effort to uplift and assist others and advanced their Progressive class interests.

Hundreds of schoolchildren from every corner of metropolitan Boston gave up their Saturday mornings to rehearse. As the musical director and "master of English pageantry" Louis W. Parker observed, these boys and girls did not engage in mere rote repetition of scenes but "entered into the spirit of the pageant in a most remarkable manner." For instance, high school girls who participated in the "spinning and quilting episodes have done the actual work over their quilting frames and spinning wheels as did their ancestors." Seized by the spirit of the Colonial revival movement, Bostonians searched their attics for "old costumes to evoke the appropriate domestic fealty."[18] For those who had only recently immigrated to Boston, this was an initiation into an imagined past, an introduction to a collective myth particularly strong in New England at that time.[19]

## THE EXPOSITION

The "1915" Boston Exposition opened on November 1 at the old Art Museum in Copley Square. The response was tremendous—an estimated two hundred thousand people attended the exhibition, and its original four-week run was extended until December 12 as a result of the enormous popular demand. The exposition was divided into fourteen sections: "City Planning"; "Labor Unions"; "Insurance"; "Housing"; "Prevention of Disease and Accidents"; "Sanitation"; "Parks and Playgrounds"; "Transportation"; "Education"; "Civil Organizations"; "Settlement Work"; "Charities"; "Work with Youth"; and "Public Utilities." Hine's thematic contribution to the exposition was a photographic investigation of the city peopled with child laborers and child welfare advocates. His photographs were exhibited in Room 30, which was dedicated to "Boys' and Children's Work" and featured exhibits by organizations with similar purpose; in addition to the Massachusetts State Child Labor Committee, which sponsored the photographs, these organizations included the Boys' Institute of Industry, the Boston Children's Aid Society, the Massachusetts Society for the Prevention of Cruelty to Children, the Boston Children's Friend Society, the Children's Mission, the Roxbury Home for Children and Aged Women, the Massachusetts Infant Asylum, the Bunker Hill Boys' Club Association of Charlestown, the West End House, and the Newsboys' Club.[20] It isn't clear from the available evidence how much planning went on between these organizations, but Hine's photographs spoke directly to their mutual concerns.[21]

In his work Hine did not simply expose a social problem and leave it to viewers' imaginations to resolve it—he also showed that something could be done to improve the situation. In the "1915" exhibit, Hine made certain to show the institutions of uplift already in the public's midst. He included two photographs to illustrate what could be done to keep newsies off the streets. One shows the inside of the Newsboys Reading Room, with boys occupying every table while a man and a woman appear to be supervising the scene.[22] It is an image of security, showing boys safely engaged in structured play. Hine's caption tells viewers that the boys are playing games rather than reading. This play is not random or haphazard, however, but well organized and well ordered. These boys were not left to wander the streets and get themselves into trouble; the

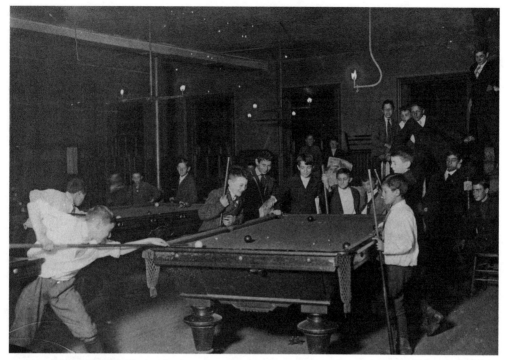

FIGURE 4.2. "Newsboy Club." (Hine 0950)

reading room served as an alternative, acceptable, and supervised space for creating community among the newsies.

Hine again captures a group of boys at play in a second photograph taken at the Newsboys' Club (fig. 4.2). In this photo the boys are playing pool, with no sign of adult supervision. The absence of an adult figure might be compensated for by the boys' being at the club and not roaming the streets. They seem to be having a good time. No money appears on the pool table, so we may assume that no wagers were placed on the outcome of the game. Five of the boys around the table hold pool cues, making it unclear if they are playing an organized game or if they are playing in teams. We can't tell whether it is night or day, though the interior of the pool hall is dark and lit by a few gas lamps, suggesting that it is evening. The overall sense among reformers in this era was that institutions, even those drawn from modest resources, protected youth and put them on the right track and out of trouble. We might question the moral landscape of the poolroom, but compared to public pool halls, this place appeared a haven from transgression. Better to supervise such spaces as this than to let boys pick up the game somewhere else.

The exhibit included several photographs of children working the streets on their own terms—peddling fruit, tending a cart in the market, running errands, or scavenging. Hine photographed these children engaged in work that is normally invisible in the statistics and in the descriptions of child labor from the period. Their labor is rarely recognized in labor history, except by those looking at family labor systems.[23] Hine does not laud the children's entrepreneurial spirit, but neither is he condemning in his captions—with one exception, where he gives just enough detail in his captioning to raise concern among viewers (fig. 4.3). The photograph shows young girls working in a street market; Hine's caption tells us that it is eleven o'clock on a Saturday night. We see one girl with her back turned to Hine, sitting at a cart owned by Antonucci, license no. 726; a bag is placed on the cart, but there's no evidence of what she is selling. Other than a woman whose face peeps out from the rear of the crowd, the young girl is surrounded by men. A police officer is at left in the image, assuring the viewer that as he stands by, the girl is safe, but should he walk away she is alone. Her father might be standing beside her, but we can't know that for sure. Hine doesn't condemn the behavior of his subject outright; he leaves viewers to make their own judgments from the evidence he presents them.

Another of Hine's photographs shows two boys in a horse and wagon carrying

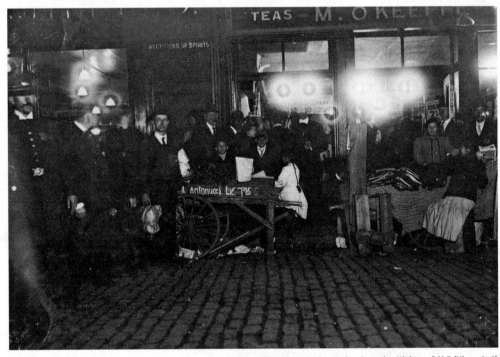

FIGURE 4.3. "Little Girls working in the Market Saturday Night, 11 P.M." (Hine 0898)

FIGURE 4.4. "Selling Grapes Saturday Night in the Market." (Hine 0911)

grapes for sale in the market on Saturday night (fig. 4.4). Underneath the wagon, a woman's dress is seen skirting forward, giving the otherwise stationary image some subtle movement. This is no fly-by-night, spur-of-the-moment activity: these boys have an operation they are managing to oversee, and the resources at their command are visual evidence of their success. We don't see hardship in the boys' faces or bodies. They appear well enough on their own.

Hine photographed another boy "in business for himself" (fig. 4.5). Again, the image may call attention to the boy's youthfulness, but it does not appear that he is suffering or threatened by any harm, injury, or danger. Two men are walking away, while another man seems to be shopping at the boy's cart, which is laden with heaps of produce. Underneath the pushcart is a large box of potatoes. The boy is rolling up a purchase for one of his customers and looks prosperous and self-sufficient, though he probably has to climb on a wooden box to reach some of the produce. Clearly these were not idle hands doing the devil's work.

Although it received some attention from the National Child Labor Committee, which produced a pamphlet on the topic, very few reformers spoke of the common practice of scavenging a city to gather wood, coal, food, or whatever else might be

FIGURE 4.5. "In business for himself." (Hine 0935)

found in streets, alleys, and dumps. In Boston, Hine produced numerous images of boys and girls scavenging, including several photographs of "woodpickers," children who traversed the city gathering up what other people had thrown away or left unclaimed. In the woodpicker image Hine included in the exhibit (fig. 4.6), a boy rests from his labors while his makeshift cart, loaded with wood, is parked in the middle of the street. Men and women pass by without looking at the cart or the boy; no one seems to give him a second thought. A police officer coming up the street shows no alarm either. This supports the idea that the woodpicker was a staple of the urban terrain.

Two of Hine's photographs of scavengers, however, do generate greater unease. While picking wood from the streets might seem to have been acceptable, if not useful, where and what scavengers collected mattered in the judgment of others. Some of the most powerful photographs Hine took for the Boston 1915 exhibit were of children scavenging and playing at the dumps (fig. 4.7). In several cities in the Northeast that Hine photographed in—Providence, Pawtucket, Lawrence, and New Bedford, for instance, as well as Boston—he discovered boys and girls working the dumps. The prevalence of this activity might demonstrate the utility of scavenging to family

FIGURE 4.6. "Boy Woodpicker Resting." (Hine 0933)

survival. A child who brought home valuable wood or goodies found at the dump helped the household save money. The items were no doubt welcome, but anyone concerned with children's welfare would obviously have found digging through the dump unhealthful. Another scavenging photograph charged with pathos of a sort shows a boy "patching up a meal" as he digs through a barrel of rubbish and waste (fig. 4.8). A man and a woman appear to have walked by the child without stopping. The boy seems to have found something to eat, perhaps satisfying some of his hunger. Images of children digging through the city trash for food generated concern and evoked shame that such conditions existed.

In a few of his Boston photographs, Hine created a sense that children were engaged in a constant effort to get something for nothing. Hine photographed three boys standing beside a fish peddler and his cart (fig. 4.9). In the background are two police officers, identifiable by their hats and accentuated by their broad bodies. All but the fish peddler and a couple of shoppers have their backs toward Hine. In the front and center of the image, the boys have arms full of wood scraps and wood boxes. Hine captioned this photograph "Swiping Behind the Cop's Back," though the visual evidence doesn't seem to support that description. If the boys are swiping something, they are very casual about it. The boy at left appears to see the policemen a few yards away, and at first glance it might look as if he is trying to pull his friend away so they

FIGURE 4.7. "Boys and Girls Working on the Dumps." (Hine 0906)

FIGURE 4.8. "Patching Up a Meal. Boston Slums." (Hine 0918)

FIGURE 4.9. "Swiping Behind the Cop's Back." (Hine 0873)

can run before the police turn around; however, although it is difficult to see, he is actually at rest, sitting on a wood piece. The boy closest to the cart seems to be asking for a box rather than taking one, or at least the fish peddler doesn't show any alarm. The caption, though, suggests a dramatic narrative of street urchins succumbing to crime to accompany a masterfully framed photograph in which Hine sketches his distinctly urban landscape. Many of the photographs Hine took for the Boston exhibit are, like this one, clearly urban, if not so clearly of Boston. He used all the signifiers of the city, from cobblestones, partially torn up in the distance for the laying of new pipe, to a warehouse sign hanging over the tough Irish cops on the beat, the immigrant fish peddler, and the three young boys, unsupervised and seemingly up to no good.

Many of Hine's photographs depict entrepreneurial children looking for opportunity, exploring ways to navigate the city and draw supplies, if not sustenance from the refuse left in the wake of modern urban life. The children, unsupervised on the city streets, show their ingenuity in countless ways. One small boy dragging a stick of lumber has made the work easier by using rope to pull the wood along (fig. 4.10). The boy seems to have realized that it would be much easier to bring the wood home if he were to hitch his fortunes to a passing horse and wagon, so he chases the wagon down

FIGURE 4.10. "Mister, Give me a Lift?" (Hine 0874)

the cobblestone street, trying to tie his rope to it. Hine's caption gives the boy words: "Mister, Give me a Lift?" Whether he actually spoke them or not, we don't know. The wagon driver doesn't seem aware of the boy. This was reflective of the lives of children in the streets: they were hard at work everywhere in the city, to the point that they became invisible in the landscape, until Hine photographed their activities and made them visible to reformers. By exposing that slice of time in the boy's day, Hine transformed the boy's labor into a valuable and long-lasting call for reform.

Hine produced several images of boys in various stages of "flipping a car," or hopping aboard the back of a moving trolley. These photographs show boys hustling and getting something for nothing—a ride here or there. The real concern, however, was the danger posed by the way the boys used the trolleys. In the photograph Hine displayed in the Boston 1915 exhibit (fig. 4.11), it appears that the boy is climbing aboard and entering the trolley as any passenger might. In other images Hine made of boys flipping cars, however, the boys are in far more compromising positions—hanging on to bumpers, riding just above the trolley tracks, seeming to court catastrophe with their casual bravado. For this image, Hine relies on the caption to elevate concern among viewers: "Sure, I take Cars!"—although it is questionable whether the boy flipping the

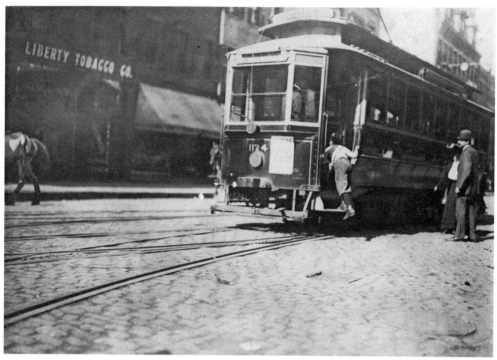

FIGURE 4.11. "Sure, I take Cars! (Dangerous.)" (Hine 0890)

car actually spoke those words. In case the image alone does not convince the viewer of the potential peril, Hine includes a reminder in his caption: "(Dangerous.)."[24] The boys made play of work, while Hine, linking words to image, decried the dangers abundant in the city streets.

If Bostonians were concerned about what children were doing, they were equally concerned with what children were *not* doing—going to school. Massachusetts had long recognized the intimate balance between going to work and attending school, and almost all early child labor legislation assigned truant officers responsibility for enforcing the legislation. Compulsory education legislation was the flip side of child labor legislation. Truant officers were generally burdened beyond just locating truants and sending them back to school; they also were responsible for investigating violations of school attendance laws and bringing charges. Boston probably had more resources dedicated to truancy than did most other cities, but still the social problem remained, along with the question of how child labor legislation would be enforced. Hine's captions identify the subjects of a few of his Boston photographs as truants. One photo he included in the exhibit shows two young boys sitting on the deck of a fishing boat that is tied up amid a flotilla of similar craft (fig. 4.12). In many respects it is a

FIGURE 4.12. "Fisher Boys Playing Truant on their Father's Smacks." (Hine 0903)

typical Hine portrait: the boys are situated in the center of the image, looking straight into the camera and into the eyes of the viewer, who connects with the boys. Hine captioned them as "Fisher Boys Playing Truant on their Father's Smacks." Was the father complicit in keeping the boys out of school? Were the boys working with the father? Going out fishing? Cleaning the deck while the boat was docked? We don't know. We do know that truancy was a component of the child labor problem, and child labor was a complex social issue requiring more sophisticated responses from people in the child-helping professions.[25]

Having put before the public a sampling of child labor activities in Boston, Hine turned to documenting some of the gains being made in children's education, labor, and general welfare. Education had always been valued in Boston, but in the future Boston of 1915, new approaches to education would be essential, given the new circumstances and demographics. Immigrants arriving in New England from eastern and southern Europe in the late nineteenth and early twentieth centuries brought their own cultures and customs, their own languages and religions. Immigrant adults and children were often left to their own worlds, spending much of their time among people who shared their culture, language, and values. Schools were arenas that brought

FIGURE 4.13. "Working Girls of all Nationalities Making the Best of the Spare Evening Hours." (Hine 0884)

FIGURE 4.14. "Working Boys of all Nationalities." (Hine 0885)

together the foreign- and native-born, where children of immigrants adopted and adapted to their new nation's culture so that they could serve within their households as the cultural bridge between the Old and New Worlds. Many reformers worried that immigrant children who bypassed school for work didn't have adequate contact with Americanizing influences, leading reformers to place a premium on evening schools for immigrant children who worked. To illustrate that emphasis, Hine created two mirror images, one of working girls in school and one of working boys (figs. 4.13 and 4.14). The perspective was the same in each classroom. Boys and girls of "all nationalities" made the "best of the spare evening hours" by going to school to learn English and improve themselves after their long days of work. It is hard not to be impressed by the children in these well-ordered classrooms. Supervised study in the evenings not only Americanized these children but also kept them from finding themselves in dead-end jobs down the road.

Many children who left school for work did so because they found school to be boring and unrelated to their working-class lives. In response, vocational training became an increasingly popular educational alternative. Vocational training strove to keep children in school longer and to give them a reason to be there by teaching them useful skills or, better yet, a useful trade. In an exhibit pointing toward a better Boston, it was important that reformers share their ideas about vocational training with the broad public in hopes of generating sympathy and support.

Vocational training was hardly a gender-neutral proposition, and Hine's photographs of vocational settings for the Boston 1915 exhibit showed the limitations in contemporary thinking regarding vocational training for women. Although many Progressive women reformers were engaged in creating new occupations in social work, child welfare, and public health, the images Hine selected to point to the future Boston did not show the "New Woman" of turn-of-the-century feminist ideology but rather were decidedly conventional in their presentation of occupational opportunity. For example, Hine photographed a group of girls who are gathered around a table in white smocks and caps (fig. 4.15). The image could be taken for one of young women conducting an experimental investigation into work hazards or nutrition, but in actuality they are, as Hine tells his viewers, "bread making as mother makes it." Ironically, the female reformers who were crafting new lives for themselves were confining the girls' education to traditional women's roles.

Another of Hine's images shows a busy classroom of "working girls learning dressmaking in the free evening school" (fig. 4.16). For seamstresses, wives, and mothers, dressmaking was a valuable skill. In the 1910s, many newspapers regularly printed dress or shirt patterns that reflected the latest fashions, making them available to the appreciative young women who couldn't yet afford ready-made clothes. Useful though it was, however, dressmaking did not necessarily provide the young women with any

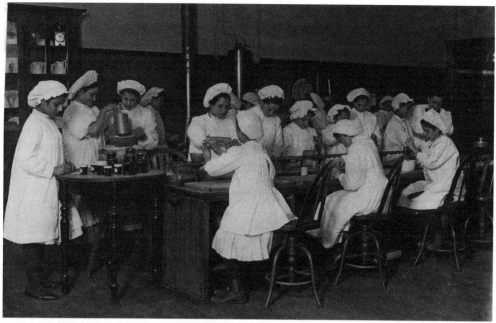

FIGURE 4.15. "Bread Making as Mother Makes it." (Hine 0878)

greater prospects for future types of employment. The emphasis in women's vocational training was on order, utility, and conformity to assigned gender roles. Again, there was no expansion in the prescribed female occupations. Rather than teaching young women administrative or professional skills, vocational training for women largely meant restricting their opportunities.

It was different for young boys, who had much broader employment and training opportunities and who were being trained not only to fill traditional occupations but also to assume a traditional role as family breadwinner. A photograph that Hine captioned "Learning a Trade" shows a group of boys standing attentively at desks and tables, focused on a task (fig. 4.17). This exemplified child labor reformers' vision of how children should proceed into the workplace. Vocational training removed many of the concerns that reformers had about children working, while showing recognition that a large segment of youth would not be going on to high school. Vocational students were being trained for skilled positions in occupations that appeared to have a future, and while they were in school, they would not be competing against their parents and siblings in the labor force, driving wages downward (fig. 4.18). Vocational training was a far more rational and efficient approach to children entering the labor market. In this way, the nation would not be grinding its seed corn, as Hine expressed it in an exhibit poster (see fig. 4.1).

FIGURE 4.16. "Working Girls Learning Dressmaking." (Hine 0883)

FIGURE 4.17. "Learning a Trade." (Hine 0892)

FIGURE 4.18. "Headed for Industry." (Hine 0893)

For his exhibit in the "1915" Boston Exposition, Hine produced photographs and posters that showed the problems present in the city, but he also documented efforts to address those social problems and to create viable paths to social health and well-being. Hine's work was ideal for the type of program set by the 1915 movement, illustrating a Boston in which children's needs were addressed through varied social institutions that worked together to coordinate limited resources and maximize social capital and political will in the city. One image in particular reflected the optimism that characterized the 1915 movement. It was a photograph of "immigrant children learning to play on the roof garden of the Washington School" (fig. 4.19). Although they no doubt speak different languages, the children have joined hands and dance in a large circle. The image reminds the viewer of what children *should* be doing: playing, and growing healthy and strong, and being joyful for Boston's more benevolent future in 1915.

## THE HARVARD SOCIAL MUSEUM

Photography was a vibrant and essential tool for portraying social conditions and enabling reformers to understand complex social problems such as child labor.

FIGURE 4.19. "Immigrant Children Learning to Play on the Roof Garden." (Hine 0946)

Reformers also used photographs in the process of determining the best course of action to address a particular problem. Photography could take a specific moment in time and freeze it forever, enabling viewers to gain some basic understanding of the social context of the image. Reformers well understood that the more snapshots they had of a problem, the better their analysis and response. Photographs were passed between organizations, recycled from one exhibit to another, and compared with each other. They were used to expand local knowledge through their deployment in exhibits, professional journals and magazines, municipal organizations, and newly emerging professional associations. A new institution for disbursing this knowledge emerged in both the United States and Europe in the early part of the twentieth century—the social museum.

The social museum was designed to approach societal problems on a more efficient, rational, and wide-ranging basis. American social museums, like their European counterparts, soon became clearinghouses for exhibition components: while some organizations kept and reused exhibit materials after the exhibits ended, others passed them along to social museums. Hine's work on the Pittsburgh Survey, for instance, was deposited in the Harvard Social Museum.[26] Similarly, materials created for a civic exposition in St. Louis were later used briefly by the Municipal Museum of Chicago before being deposited at the Chicago School of Civics Social Museum. Maps, charts, placards, banners, screens, objects, models, pictures, photographs, enlargements, blueprints, lantern slides, motion films, and printed ephemeral material from previous exhibits were sorted, arranged, and indexed by social museums and made available for organizations to reconstruct in new combinations for fresh exhibits. Social museums often also housed specialized exhibition equipment such as map cases, card exhibit bases, vertical files, label cases, book stacks, exhibit display frames, lantern slide projectors and cases, and hanging chart cabinets.[27]

In 1903, Professor Francis Greenwood Peabody established the Social Museum at Harvard University as part of the Department of Social Ethics to provide a more rational and scientific approach to the social question and to direct the amelioration of industrial and social life. The museum housed extensive collections of materials, particularly graphic materials, illustrating social settlements, housing and city planning, industrial conditions, social surveys, and so forth. The assumption behind the museum was that "the most immediate need of students" engaging the social question was "not merely enthusiasm or sympathy or self-sacrifice or money, but wisdom, discretion, the scientific interpretation and comparison of facts." University teaching, according to Peabody, was "being revolutionized by new applications of the inductive method." Application of the inductive method was encouraged by "setting before the student in graphical illustrations the evidences of progress in various countries and putting at his command the fund of experience accumulated in various parts of the world." The museum was a centralizing institution, connecting reform efforts and ideas useful to social reform.

Peabody argued for collecting evidence broadly and approaching issues from a comparative perspective. The "remoteness" of the United States from other countries and the "brevity of its social history" made it important that illustrative material from beyond the United States be provided to students and academic researchers. For instance, Peabody claimed that "Germany has much to teach the United States of municipal administration, but may learn much from America concerning the free conciliation of labor-disputes, or the science of improved dwellings." England could teach Americans something of trade unionism and industrial cooperation but would have to "turn to the United States for lessons in the reform of the drink-traffic." Comparative analysis of social problems was vital to constructive work in the same way that comparative zoology provided the naturalist with a corrective against "hasty judgments" and served as the "prerequisite of judicious conclusions."[28] Comparative sociology adhered to professional standards and practices and was self-consciously nonpartisan.

Writing in 1908, Professor Peabody claimed that many "notable" institutions in countries such as Germany, France, Holland, Austria, and Spain had been established in recent years to promote social welfare through the exhibition of comparative results. New York's Institute of Social Service and its associate, the British Institute of Social Service in London, both collected "all manner of facts bearing on human progress," making the experiences of all countries and individuals available for guidance. What made Harvard's Social Museum stand out from these other collections was that none of them, despite their scope and intention, were associated with academic life or "primarily concerned" with the instruction of university students. Harvard's Social Museum, Peabody maintained, was "the first attempt to collect the social experience of the world as material for university teaching, and to provide guidance for academic inquirers into the study of social progress." The social question, "seen in the mirror of a social museum, reveals its essential character as an ethical question. . . . The facts collected in such a museum are dead material until touched to life by the interpretive power of philosophy."[29] The museum gave life to Lewis Hine's photographs and, in their display, kept the images alive in the reform discourse.

# CHAPTER 5

# Sardines

———

B Y 1900, ACCORDING TO THE historian Jane E. Radcliffe, Maine was the United
States' only producer of canned sardines. Goods such as corn, blueberries, and
lobsters were also canned in the state, and Maine's canning industry was second only
to the textile mills in the employment of children. Like other forms of rural employment, such as logging, fishing, trapping, and farming, canning was a seasonal industry.
Labor was irregular, and the season was short, generally lasting from early May to
September. While the season was on, men, women, and children put in long hours
of intense work in the canneries. Like many other forms of agricultural labor, canning was frequently family-based, with all members contributing to the household
income. The industry was well suited to children's labor, as it required large numbers
of unskilled workers who needed little in the way of previous training. Children were
easy to teach and of limited expense to the operators. The seasonal quality of the work,
necessitating a casual reserve army of labor, was also well suited to the employment of
women and children, who could be hired and laid off without serious disruptions to
the rural economy, where additional income was always welcome and supplemental
occupations such as logging, trapping, and fishing were common.[1]

## STATE EXEMPTIONS TO CHILD LABOR LAWS

Maine's child labor laws exempted children engaged in work in "any manufacturing
establishment or business, the materials of which are perishable and require immediate
labor thereon, to prevent decay thereof or damage thereto."[2] This exemption negated
what modest protections children working in manufacturing enjoyed and, because

it was extended to the canning industries, gave the canning companies access to a pool of labor without restriction. The exact number of children who were employed in the sardine canning industry was difficult to determine because of the seasonal nature of the business. Children frequently received "checks" or "scrip" for the specific work they did, and often women, children, and other seasonal workers were left off the books altogether. The unpredictability of the fish catch and availability of fish for processing led to further irregularity and fluctuating demand for labor. The decennial gathering of statistics in the U.S. census failed to account for these workers, because the census was conducted during the canneries' off-season, and many of those who labored in the Passamaquoddy Bay region came from other towns to work and then returned home at season's end.[3]

Ernest Stagg Whitin, writing in the *Outlook* in January 1905 about the lax enforcement of New York State's child labor laws, found in the canning industry that "most of the factories are provided with long rows of benches on which sit the women surrounded by their children. . . . Half are usually young children under fourteen years." Children worked from nine in the morning until nine at night, and to avoid loss of time their lunch was brought to them, to be eaten in the factory. Whitin appealed to the canning industry to "substitute machines to do the work of the little hands." This change had occurred in other industries, and when factories had been forced to introduce machines to replace child workers, they had "found that they could make a profit against the unfair competition of even the employers of cheap child labor."[4] However, as long as the perishable food exemption in Maine's child labor laws remained operative, canning companies in that state saw no urgency to invest in machines to replace their child workers.

John Spargo's sociological classic *The Bitter Cry of the Children* (1906) attracted particular attention to Maine's sardine canning industry and its detrimental effect on educational opportunities for children. Spargo reported that in 1900 there were 117 plants in Maine engaged in the preservation and canning of fish—specifically, canning small herring and marketing them as sardines. The industry was centered in Passamaquoddy Bay, in the coastal towns of Lubec and Eastport in Washington County. Spargo admitted that he had not visited the area, but his impression from speaking with "competent and trustworthy sources" was "that child slavery nowhere assumes a worse form than in the 'sardine' canneries of Maine." Spargo quoted at length from a private letter written to him by one of his sources:

> In the rush season, fathers, mothers, older children, and babies work from early morn till night—from dawn to dark, in fact. You will scarcely believe me, perhaps, when I say "and babies," but it is literally true. I've seen them in the present season, no more than four or five years old, working hard and beaten when they lagged. As you may suppose, being out here, far away from the centre of the state, we are

not much troubled by factory inspection. I have read about the conditions in the Southern mills, but nothing I have read equals for sheer brutality what I see right here in Washington County.[5]

Spargo's charges stimulated vigorous debate in Maine, leading to a call for new and more effective child labor legislation. In his annual address to the state legislature in 1907, Maine governor William T. Cobb called for improvement in the state's child labor laws. He appealed to the state's legislators to be the children's "champions," arguing that "neither the thoughtlessness of parents nor the indifference of employers must be permitted to interfere with the performance of the State's manifest duty to provide, as best she may, for the moral, physical and educational welfare of these children to whom unfortunately so many of the pleasures and opportunities of childhood are denied." For children laboring in industrial occupations or the street trades, the state was empowered to intervene and play the role of guardian between the child and the market, but children working in agriculture, particularly the canning industry, remained outside government protection. The Maine legislature revised the state's labor laws in 1907, but it retained Section 55, the "perishable goods" exemption.[6] Thus, despite legislative action, children in the sardine industry continued to be largely unprotected.

In an effort to test and refute the veracity of Spargo's claims, Maine's Bureau of Industrial and Labor Statistics initiated an official investigation of the canning industry in Washington County. Eva L. Shorey of Bridgton, Maine, a special agent of the bureau, was instructed to "make a full and thorough investigation of all existing conditions in connection with the industry, and to ascertain so far as was possible any facts that would substantiate or contradict the serious charges that had been made."[7] Shorey found conditions in the canning industry to be less than desirable but not as bad as Spargo had charged. In her report she noted the "repulsive" aspects of the work done by "women [and] small boys and girls" in the canning factories and the long working hours.[8] She also acknowledged that "children under fourteen are employed in some of the factories" and lamented the "intermittent" education those children received as a result of being "out of school so much of the time."[9] However, she stopped short of making recommendations for stepping up the enforcement of school attendance laws; and while her report mentioned the "perishable goods" exemption in passing, it offered no opinion on the reasonableness of that exemption being extended to the fish canning industry.

Shorey rather defensively denied that the sardine canneries were "slave driving" operations. On the contrary, she noted, "young children come and go as they wish. It may not be very attractive or desirable work for one of tender years, but it is honest and healthy and does not continue day in and day out for any great length of time consecutively. The children appear to enjoy it and are very proud to tell how many boxes they have cut."[10] Shorey noted that some of the young children wanted to work and were in no way compelled by their parents to do so. She described one instance in which she came upon a six-year-old girl washing her hands. When asked what she had been doing, the girl told Shorey that she had just cut two boxes of sardines, and she showed Shorey the ten cents in checks she had received for her labors. To reinforce her point that this labor was not only voluntary but also desirable on the part of children, Shorey noted that she "later saw the young child enjoying a long stick of striped candy, which her earnings had provided." The girl was from a nearby town and lived in the camps only during the sardine season, and she told Shorey that she "worked with her mother when she wanted to."[11] The labor regime was loose and irregular, and children entered and exited as they and their families dictated, but when a ship came in, the call for labor went out. An 1887 report of the U.S. Commission of Fish and Fisheries described the gathering of a workforce: "When the boat nears the wharf, the cannery whistle or bell is sounded as a signal for the cutters, who are usually boys and girls from eight to fifteen years of age. These are presently seen brandishing their large knives as they rush through the street on their way to the building."[12]

Responding in 1908 to Spargo's exposé and Shorey's report, Ernest Stagg Whitin argued against the inclusion of the sardine canning industry under the perishable goods exemption to Maine's child labor laws. Whitin acknowledged that beans and other agricultural goods of that type had to be "harvested and prepared within a limited time" and thus might meet the legal criteria for "perishable foods." He argued, however, that "the proof remains with the canners of fish to sustain their claim of inherent perishability of their goods." Preventing spoilage of the fish catch was simply a matter of fishermen not overloading their boats with more fish than could "properly [be taken] care of" when they reached land—a practice that had been observed by "Maine fishermen and their ancestors since the time of the Vikings." Whitin also noted that fish was classified as meat under the nation's Pure Food Law, and so the head of the Federal Bureau of Animal Industry could be empowered to place the Maine canneries under government supervision; at that point "model conditions would be established and machinery introduced which would do away with any justification of child-labor."[13] Whitin reflected the Progressives' optimistic faith in the state: the government would supervise the relationship between child worker and employer, encouraging the replacement of children with machines and ensuring more favorable conditions for those that did labor.

## LEWIS HINE INVESTIGATES THE INDUSTRY

In August 1911, Lewis Hine journeyed to Eastport and Lubec, Maine, on behalf of the National Child Labor Committee to conduct an investigation of the sardine industry and to gather graphic evidence that would demonstrate the need for future legislation protecting child laborers. Eva Shorey had defensively attempted to put the best light on children's involvement in the sardine industry. Hine, like his friend John Spargo, attempted to bring about reform. Hine produced forty-five photographs with the usual captioning and submitted a report on his fieldwork to the NCLC.

Hine viewed Eastport and Lubec as "one industry" villages in which all local resources were centered on canning. One of the chief drawbacks to the industry, according to Hine, was that the "whole year's work" was "concentrated into five months, from May to September." Workers in these canning villages had no other work to do, unless "they moved away," and therefore had to derive their entire year's income from a few months of highly intensive labor, leaving many of "them idle for eight or nine months." Other "disadvantages" intrinsic to the industry were the "fluctuations in the movement of sardines caught," which made work irregular and intensified labor "on certain days and nights during those few months."[14] This was casual labor, gathered at the sound of the horn and dispersed at the end of handling the haul.

As Hine worked his way through the community, he came to conclude that conditions in the industry had changed in recent years. The sardine companies continued to increase productivity while also reducing the price of sardines. Hine reported that the output of Eastport and Lubec amounted to "over one and a half million cases of sardines annually." But a case that once retailed for $10 or $12, according to Hine, was now $2. Unfortunately, the increase in quantity that enabled the drop in prices came at the cost of decreasing quality. One advantageous change in the industry had been the dramatic reduction in the amount of handwork involved, largely because the "covers of the sardine cans were now pressed on by machinery instead of being soldered." Hine believed that the cans were "not so air tight," and the fish were increasingly "canned with the entrails, the head only being snipped off." Hine undoubtedly observed this drop-off in quality control himself or learned of it from several of the cutters and canners he photographed.

The increase in productivity and decrease in retail prices created a constant pressure on factories to intensify the labor regime and to find cheaper sources of labor. Workers' annual earnings declined through overall reductions in wages. Hine reported that men who had once earned between $25 and $40 per week were receiving only $15 to $20 a week in the summer of 1911. Along with the constant downward pressure on wages, many men had "been replaced by machinery and by women."[15] Women were "sweated," as were the children who worked in the industry. Rather than question the underlying assumptions that consented to female exploitation, Hine rationalized the labor system, reporting that "men do not seem to be so well fitted to packing as women," who, accordingly, "make much more than the men when they work overtime, nights and Sundays, as they often do." Declining wages, one of the "bosses" told Hine, meant that "if a man wants to get along here, he wants to have a wife and a dozen kids all working, and then he'll have to scrape to get along." On the other hand, children's labor had been reduced by eliminating or changing stages in the work process. Hine noted "less child labor at present than there used to be; chiefly because the cutting off of the heads of the fish by the children before the fish are baked has been greatly reduced and in some places eliminated by the practice of having the packers snip the heads off with scissors as they pack them."[16]

Children's presence within the industry was not eliminated entirely, however. Hine photographed numerous young children, mostly boys, cutting fish (fig. 5.1). Hine described the work of cutters, who "cut the heads off while the fish are raw (some do this to the larger fish only, and some not at all). . . . The fish are ladled out on the cutting tables which accommodate eight or ten workers." Cutters picked up the herring in their left hand and, with the "cutting or chopping motion of a great butcher knife," severed the heads, dropping the bodies into a wooden box, which, upon filling, netted the worker five cents. The smaller children, he observed, handled only one fish at a time at first, but as they became "more proficient" they were able to cut two or three at once.[17] One of Hine's photographs shows a young cutter working at a table that is almost as high as his chest (fig. 5.2). In this and several other images, Hine placed men at the cutting tables next to the children to juxtapose their ages and sizes. Was cutting a man's task performed by boys? Were the men supervising or driving the young workers? Were these coworkers under the same clock? Hine displayed these young boys and, less frequently, girls wielding their "large, sharp knives," posing an imminent danger to themselves and those around them. A child who looked away for one moment might slice open a finger or even lose a finger entirely. Hine's caption notes that "the slippery floors and benches, and careless bumping into each other increase the liability to accident." We know the workers all have cuts because Hine recorded their complaint that "the salt gits in the cuts an' they ache." A viewer reading this complaint back in Hine's time would perhaps have wondered whether these painful and dangerous labors should be deemed hazardous and regulated on those terms. Could a machine be designed to perform such work?

Hine's photograph of three cutters in the Seacoast Canning Company factory who "work regularly whenever there are fish" makes clear that they were serious and productive workers, even at their young age (fig. 5.3). The three boys—seven-year-old Grayson Forsythe on the right, nine-year-old George Goodell at center, and George's brother Clarence, a barefoot six-year-old who "helps [his] brother"—stand proudly in front of a huge heap of fish heads that testifies to their labor value. In Hine's caption for his photograph of nine-year-old Hiram Pulk (fig. 5.4), who "cuts some" in the Seacoast factory, Hine quotes the boy as estimating his own value within the workplace: Hiram acknowledges that he "aint very fast—only about five boxes a day," for which he received a total of twenty-five cents. His value as a photographic subject for child labor reform was higher. The small boy looks directly at the camera, somewhat

FIGURE 5.1. "Three young cutters who work in Seacoast Canning Co." (Hine 2450)

FIGURE 5.2. "Shows the way they cut the fish in sardine canneries." (Hine 2435)

FIGURE 5.3. "Three cutters in Factory #7, Seacoast Canning Co." (Hine 2421)

FIGURE 5.4. "Hiram Pulk, 9 years old, cuts some in Seacoast Canning Co." (Hine 2427)

forlorn, appealing to the photographer or the viewer for empathy. As usual, Hine has positioned his subject in the center of the image, the piling and wood frame and the sardines and cutting around him in sharp focus, while the bay behind Hiram is a blurry backdrop that brings the boy into greater visual relief. In this very traditional work portrait, Hine had Hiram pose with a fish in his hand. He holds the fish so it is up against his body, which is filthy, his shirt hardened by fish guts and never quite dry, his bare feet and thin grimy legs evidence of his labors and hardships.

Another of the several traditional types of work portraits that Hine created was that of the worker with his or her tools in hand. In the canning of sardines, the knife was the tool of choice, and Hine made good use of it both as a prop and as an immediately recognizable representation of the dangers posed to industry children. Hine photographed Ralph, a young cutter, leaning against a piling (for style, not support) and holding a large knife in his left hand (fig. 5.5). Ralph is playing every bit the part of a cutter; with his bib and overalls tied at the waist, his pose cool and relaxed, he looks like an old hand. He holds the knife casually, his fingers bandaged from cuts. Hine's caption notes that cut fingers were ubiquitous in the trade, and "even the adults said they could not help cutting themselves." That children were cutting themselves as a matter of course was an issue of concern for reformers but also an opportunity in their quest to generate a sympathetic audience.

The combination of children, knives, and slippery floors made cutting a very dangerous job. The work was "simple," Hine noted, but also hazardous. The floors were constantly "wet and slippery with brine. Children stood all day with wet feet, handling cold fish." The shacks were open to all sorts of "inclement" weather. Hine reported that "slippery floors, slimy fish, personal carelessness, and jostling of neighbors combine to make gashed fingers common," and although the gashes were usually not serious, they could become so without attention. The salt water that the workers toiled in further aggravated their cuts. As Hine documented so well in his photographs, "nearly every child has scars of fresh cuts to exhibit." One man said to him, "You can tell a cutter by the notches on his hands."[18]

Whenever Hine came across a child he believed embodied the injustice of child labor, he tended to take multiple photographs until he found the image that would evoke the greatest response. In Eastport, Hine met Minnie Thomas and photographed her at least four times as she performed different tasks. For one photograph Hine created a traditional work portrait of Minnie, posing her with a sardine in each hand, her huge knife wedged into a wooden fence that she stands behind (fig. 5.6). One is left to wonder how nine-year-old Minnie was capable of wielding this "average size" knife, which is about a third of her size and is a bit intimidating. The knife wasn't a tool that any nine-year-old should have been using. Hine's caption tells us that Minnie worked regularly at Seacoast as both a cutter and a packer, though mostly as a packer, and that

FIGURE 5.5. "Butcher knife used by Ralph, a young cutter in a canning company." (Hine 2446)

FIGURE 5.6. "Minnie Thomas, 9 years old, showing average size of sardine knife used in cutting." (Hine 2439)

FIGURE 5.7. "Packing sardines in Seacoast Canning Co." (Hine 2440)

she sometimes worked in the evenings when a big load of fish had arrived and the factory was busy trying to process the haul.

Hine's attempt to get Minnie to pose for him on another occasion was apparently rebuffed by her "Boss," who "would not let nine year old Minnie pose." Why he refused Hine's request is unclear—he may have thought it would interrupt her work, or perhaps he recognized Hine's purpose. Unable to capture the image he wanted, Hine took the best photograph that he could (fig. 5.7) and made certain to caption that Minnie "works with her mother, packing even far into the night"—another example of Hine relying on the verbal when the visual did not live up to its promise. The image was nonetheless very interesting, one of the few photographs taken by Hine that show the inside of the packing sheds.

A third image of Minnie shows her inside a cutting shed (fig. 5.8). Eight-year-old Clarence (see fig. 5.3) works at her side. The perspective in this image is unique among the photographs in this investigation. Hine photographed the two young cutters from a low angle, his camera even with the flooring. The viewer sees the children's backs, the table they work on, and the adults across from them who share their workspace. Only because of the angle of the photograph does Minnie's head seem to come to the same height as the heads of the adults across from her. In front of the workers are the boxes they must fill for their five cents. Hine, unsure that viewers would understand the full

FIGURE 5.8. "Interior of a cutting shed in Maine." (Hine 2422)

import of the scene, notes in his caption that the image does not show "the salt water in which they often stand" nor "the refuse they handle." By including these details in his caption, Hine sought to make Minnie the focus of sympathy.

Hine ends his series on Minnie with a traditional frontal portrait: no tools, and nothing else to offer context except the weatherworn side of a wood building that dwarfs her. This was perhaps his simplest image of Minnie, emphasizing the girl rather than the worker. She looks uncomfortably at Hine, and thus at the viewer, perhaps questioning what she is doing there. The image does not speak alone; Hine's caption interprets it for the meaning Hine intended to impart to his audience. Hine apparently located Minnie's mother, who supplemented whatever information Minnie had given him, affirming the dangers of children using the long knives: "Some of the children cut their fingers half off," she confessed. The caption also tells us that Minnie's father and grandfather both worked in the factory with her. The mother and father stayed in Eastport in the winter, while Minnie and her aunt lived in Grand Manan and returned each year during the herring run.[19]

Hine did not often produce sequences of photographs linking unfolding action, but in the case of Phoebe Thomas, an eight-year-old Syrian girl who worked as a cutter in one of the Seacoast canning factories, Hine photographed his only action series. Hine took four photographs that are loosely linked in time but tightly woven together in subject and story. The subject of the series is Phoebe, and the story Hine tells is of the pervasiveness of workplace accidents in general and the deep cut suffered by Phoebe in particular. In the first image, Hine photographed Phoebe walking to work at 6:00 a.m. carrying a "great butcher knife, to cut sardines." Hine identifies Phoebe as a cutter whom he later saw working.[20]

The next three images continue Phoebe's story. In the second photograph (fig. 5.9), Hine captured the drama of Phoebe running home after she had cut her finger, "her hand and arm bathed in blood." She had "cut the end of her thumb nearly off" and was "crying at the top of her voice." The child was "sent home alone" while her mother continued to work. Given the "considerable" loss of blood, Hine feared the young girl might faint along the way, so he followed her. Remarkably, Hine was able to place her in the center of the image and keep her in sharp focus, though the colors and shades of her dress and the stones along the stairway path slightly camouflage her. There was no hiding her voice when she cried out, but Hine's audience couldn't know that unless Hine shared that chilling detail through his caption. Phoebe made it home safely, and Hine photographed her again a short while later (fig. 5.10). To his amazement, in less than a week the girl was back at work, cutting sardines (fig. 5.11).

Many children worked as cartoners in the canning industry, putting the sardine

FIGURE 5.9. "In center of the picture is Phoebe Thomas." (Hine 2444)

FIGURE 5.10. "Phoebe, a little while after the accident." (Hine 2445)

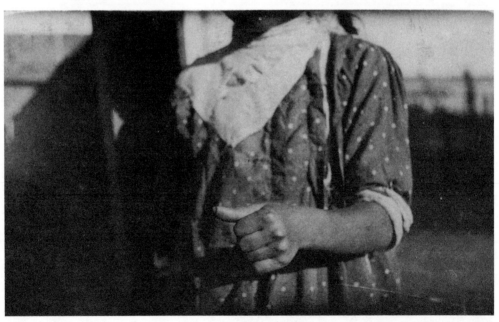

FIGURE 5.11. "Phoebe's thumb, a week after the accident." (Hine 2449)

FIGURE 5.12. "Group of 'cartoners' in Seacoast Canning Co." (Hine 2432)

cans into paper cartons and packing them into wooden boxes. The danger for these children was less immediate, but they were not free of the consequences of child labor. Cartoners were often very young. In Seacoast factory no. 7, Hine counted eleven out of fourteen children who were under twelve years old (fig. 5.12). Some of the work, he noted, was done by "adults in spare time," but much of it was done by children between seven and fourteen years of age. It was a "simple" enough process, and very easy to learn, but "very monotonous." On top of that, this part of the operation was often "speeded up in true sweatshop style." The cartoning rooms, filled with young children, nonetheless reminded Hine, in a sad and ironic way, of "kindergarten classes."[21]

In his photograph of a young cartoner named Elsie Shaw, Hine inadvertently stumbled onto a child labor issue of a different sort: children performing in the theater (fig. 5.13). Elsie's father was a boss in one of the Seacoast cutting rooms and apparently approached Hine to take photographs of his daughter. Mr. Shaw told Hine that Elsie was doing a "singing act in vaudeville in the winter" months and could use some photographs because she was "old enough now to go through the audience and sell her own photos."

Hine saw fewer children working in the packing rooms than in the factories' other departments, but the presence of children, some "even as young as 9 years," illustrated

FIGURE 5.13. "Elsie Shaw, a 6 year old cartoner during the summer." (Hine 2408)

the "fact that they can do it." Packing consisted of snipping off the heads of the sardines and packing the fish in boxes. According to Hine, bosses preferred hiring women and older girls for this work because it was "very particular work, needing care, speed, and endurance." (fig. 5.14) Men, on the other hand, "did not seem fitted to earn much at it." Although Hine found only a small number of children packing, he remarked that it was "not stretching the imagination to look forward to the day when the demand for cheapness which is pressing the canners all the time will cause them to resort to the employment of young children." Because immigrant workers were increasingly replacing the native-born in this function, Hine believed that "the likelihood of young children being drawn into that branch of work" also would increase.[22] Given the degradation of labor, it was only a matter of time before immigrant children were packing.

Wages for children working in the sardine canning industry were, as Hine stated in his report, "immaterial" and "irregular." Pay for children varied widely, and according to Hine, many of the children really did not "know how much they amounted to." Some of the youngest children earned as little as twenty-five cents a day (fig. 5.15). Some of the ten- and twelve-year-olds told Hine that they made anywhere from $1.00 to $1.50 for a hard day's work. Boys twelve to fourteen could earn $2.00 or more a day. A ten-year-old named Eugene Clark told Hine, "I made $7.86 this week cartoning."[23] The superintendent of one factory pointed to a group of women and told Hine that "one of these women made $20 last week packing." The children were vital to their

FIGURE 5.14. "Adult packers in the Lawrence Canning Co." (Hine 2437)

families' earnings, he told Hine: "When they have youngsters helping them they make $25 and up."[24] Large families endowed with many helpful hands might earn a week's living in a day's time.

In a lengthy caption for his photograph of the Hamilton family in Eastport (fig. 5.16), Hine describes the family's economic circumstances, what each family member did, and what each of them earned. The father, Hine states, was "dissatisfied with the irregular income." It isn't clear how in-depth Hine's discussion with the Hamiltons was, but Hine apparently tried to explain to the father the "connection between his early boyhood work and his present stagnation." However, the father just "couldn't see" it. Hine observes, with frustration and perhaps some disgust, that the father is "putting his little ones through the same process."

Hine also expressed concern about the morality of the canneries and the impact of such places on children. In his eyes, the "work is certainly degrading for such little ones." The "atmosphere of the slaughter house, and the attitude of the butcher" pervade the canneries, creating an environment that is "surely not good for growing children." Hine also noted that children's association "with men the type commonly found in the canneries, habituated to the use of low and profane language, is bad." In addition, Hine

FIGURE 5.15. "All these boys are cutters in the Seacoast Canning Co." (Hine 2420)

FIGURE 5.16. "Hamilton Family, sardine works, Eastport." (Hine 2453)

Figure 5.17. "Housing conditions in settlement at Seacoast Cannery #7." (Hine 2433)

feared that the irregularity of the work gave the children a false understanding of their compensation. Because they might work for a period of time, "day and night, perhaps, then loaf awhile," they were apt to "count pay by the best days, instead of yearly, and so become dissatisfied with jobs that pay less a day, but much more in the long run."

Hine photographed housing conditions that "were not the best," and he worried about their impact on children's health and welfare. During the sardine fishing season, families "crowded into temporary quarters" that were "crude and often unsanitary." Many families came from out of town and returned home at season's end. Worse off were those families who lived in their temporary housing year-round. Hine photographed the Goodell family at their home in Eastport (fig. 5.17). Mr. Goodell worked full time in a nearby mill, and during the sardine season, four or five of his children worked in the canning factories (see fig. 5.3).[25]

While in Maine Hine met with Fred Benson, the superintendent of schools for Eastport and Lubec and the only public official he spoke to who had any oversight of the child laborers. Hine reported that Benson claimed to be "interested in the regulation of child labor" but was unable to "keep very closely in touch with the problem." Child labor affected school attendance, Benson admitted, but he still maintained that attendance in Eastport and Lubec was "better than that of the average Maine towns throughout the county." Benson told Hine that he believed the factory owners in these towns were genuinely "desirous of cooperating with those wishing to regulate child labor." He even told of several owners in Lubec who had come to him to ask for the real ages of the boys. Benson had faith that the owners were "really trying to better working conditions there." At the same time, Benson sympathized with the "temptations" to use child labor "when the work presses," but he tended to place less fault on the owners than on the parents, who "need the money the children bring in." Benson put particular responsibility on mothers, arguing that women's participation in the canning industries "plays havoc with the home life," leading them to "neglect the home." Additional winter work, Benson believed, alleviated some of these problems by keeping "the people occupied" and giving the families additional income. Hine's report gave the impression that the superintendent did not have a good understanding of the scope of child labor and its impact within the district. Benson received a list of children promoted in grade level at the end of each school year and followed up at the beginning of the next school year to see that they were still in attendance. Admittedly, he had no knowledge of those children employed in the canning industry who came from other towns to work in Eastport and Lubec.[26]

Ultimately, Hine placed his hope not only on the passage and enforcement of stricter laws but also on the potential of further mechanization to diminish the use of child laborers. Hine discovered relatively few children operating or helping to run the machines that were increasingly used in the industry. He noted that a machine had just been "perfected" and tested at the Seacoast canneries that did "all the cartoning by machinery, reducing the number of workers and eliminating the children altogether." Hine believed that its success would "reduce the number of children in [those] factories" that could afford to mechanize the cartoning process. Hine did not know, however, whether initial investment in such machinery would be cost effective for all companies. Some, he suggested, "may be tempted to use children more, in order to keep down expenses" and remain competitive.[27] Sardine canning could be further mechanized, but the profits derived from the labor of little hands were too great to overlook.

Hine maintained that regulation by the state was still the most promising solution to the child labor problem. The state, he argued, already had "many restrictions regarding seining, the building of weirs, etc., to protect the fish from extermination." He acknowledged that such regulation often "caused personal hardship" and placed burdensome "limitations" on people, but it was nonetheless "necessary for the future of the industry." So, he felt, should child labor legislation be enacted and enforced to "prevent the using up of the children." Hine concluded his report with an expression of faith that if "these canners really tackle the problem of industrial conditions in this community in the same spirit with which they take up the problem of machinery, that they will find their way out."[28] For the future of the industry, mechanization would be introduced to replace casual labor with the regularity of machines.

## CHAPTER 6

# Seasonal and Family Labor

———

"Farmwork for children is the form of child labor to which we are most accustomed," argued Ruth McIntire of the publicity department for the National Child Labor Committee; "therefore it is the more difficult for us to view it as abnormal or unnecessary." According to the 1910 U.S. census, more than 2 million children between the ages of ten and sixteen were gainfully employed, and three-quarters of those children were working in agriculture. Agricultural labor went virtually unprotected by legislation, however, with the notable exception of compulsory education laws—and those laws, according to McIntire, were generally unenforced in rural districts, where employers were "protected by public ignorance" and "public sentiment that 'the farm is a good place for a child.' "[1]

Although farmwork was the most widespread type of child labor, it was the least studied and the least understood. In 1910, the National Child Labor Committee determined to rectify that. Through survey, investigation, and public exposure the NCLC hoped to "dislodge from the popular mind the tradition that all child labor in agriculture is good for the child." The committee mounted a campaign to alter that image sufficiently to secure legislation for the protection of child farmworkers. The NCLC had recently completed a survey of cranberry bogs in New Jersey and published its findings in the *Survey* (February 1910). The report was "bitterly attacked" by industry representatives. To substantiate the report's claims, the NCLC sent its investigators to survey the cranberry bogs of Massachusetts. The Massachusetts survey corroborated the report made on New Jersey cranberry workers, though wages, housing, conditions of child labor, and so forth were "far inferior" in New Jersey. The NCLC investigators in Massachusetts did find children as young as four or five years old working regularly during picking season, children deprived of the first five or six weeks of schooling, and housing conditions on many of the bogs that were "unspeakably indecent and unsanitary." The "whole situation," they concluded, "calls for careful attention."[2]

In the second decade of the twentieth century, Lewis Hine sought to give agriculture that "attention," as he had done for manufacturing and the street trades. He visited New England several times between 1911 and 1917 to survey and document different types of agricultural labor, including wage workers in the cranberry bogs of Massachusetts and the tobacco fields of Connecticut, and children laboring on family farms in Massachusetts and New York. Hine hoped that through the resulting photographs and reports he could undermine the popular notion of farm labor as being beneficial to children and replace it with greater sympathy for young farmworkers and concern for their futures.

## FAMILY FARM LABOR

Hine photographed different types of farm laborers in New England, the primary division being between those working as farm laborers for wages and those working on their family's farm. In August 1915, Hine found Jack, an eight-year-old boy laboring on his family's farm in western Massachusetts (fig. 6.1). Hine took seventeen photographs of Jack, far more than he took of any other child in his work for the NCLC. Hine's captions for these images remind the viewer that Jack was in imminent danger, although from a more gradual threat resulting from constant overwork. It was a good thing Hine captioned the photographs of Jack as he did; otherwise his audience might not have arrived at the interpretive conclusions he intended. Jack appears to be hard at work in most of the images, even impressively so, but at no time does he seem distressed or overwhelmed by the task at hand. Not only is he often smiling, but he even appears to be having a good time in his work. Looking at these photographs, one might just as easily conclude that agricultural labor was far better for the child than industrial labor or working the street trades. The dissonance between the photographs of Jack and Hine's captions for them undermines the photographs' authority and usefulness as tools for reform.

Hine captured Jack's energy and enthusiasm and documented Jack's many abilities as he photographed the boy performing a host of farm tasks. Driving a horse rake, milking cows, roping a calf, caring for a colt, carrying cans of milk on a stone boat, driving a load of hay—no matter the chore, Jack accepted the challenge and committed himself to its completion. Hine's photographs show a child who looks comfortable with his chores, so much so that even Hine has to note that "this is recreation for Jack." Jack is making play of his work. However, although Jack seems joyful, Hine cautions

FIGURE 6.1. "Eight-year old Jack on a Western Massachusetts farm." (Hine 3974)

FIGURE 6.2. "Eight-year old Jack taking care of the colt." (Hine 3972)

that "he is a type of child who is being overworked in many rural districts."[3] Yet Jack appears to be thriving as a boy growing up in rural New England.

If Hine's captions were not there to point viewers to a very different interpretation of the images, Jack might be perceived as a model young boy growing up in an idealized world of farm labor (fig. 6.2). Many people in Jack's era saw such labor as character building rather than debilitating. They believed that farmwork helped develop physical vitality and moral integrity. Farm labor was vocational in nature, fostering skills and knowledge that had future utility. The character-building regime of "useful" labor by boys like Jack helped to "preserve the solidarity of the race." The training that boys received on the farm, a scholar of the subject suggested, was "one of the best means of preserving our social democracy." Hard work, self-reliance, and spiritual connection to nature were important components in creating a "good life," and "the good life is a happy life."[4] Hine's images of Jack are positive evocations of farm life.

Hine's photographs of Jack at work leave little or no doubt as to the importance of the labor that child farmworkers performed. Jack appears to have taken on major chores in addition to milking cows and roping calves. Hine photographed Jack driving a load of hay, for instance (fig. 6.3). Jack stands atop the hay wagon with his legs spread wide for balance, his sleeves rolled up, and his arms outstretched, holding the reins. All around him are adults playing their roles in the process, but Jack stands above them and even appears to be in charge. He performs adult labor with adult coworkers at his side. There is very little that is childish about this work, however much Jack might turn it into play. Jack threatened his safety by taking on the tasks of adults, such

FIGURE 6.3. "Eight-year old Jack driving load of hay." (Hine 3973)

FIGURE 6.4. "Eight-year old Jack driving horse rake." (Hine 3961)

as riding on top the hay wagon, or working the machines on the farm (fig. 6.4). Danger was useful to Hine in marshaling sympathy for the young farmworker.

While Jack appears to be confident and in charge when using a horse rake, Lucy Saunders, whom Hine also photographed on a western Massachusetts farm, does not look as if she is building character in similar fashion (fig. 6.5). Lucy is hooking up the rig rather than driving it. We are not warned of any danger but are merely shown the task of hitching a team to the horse rake. The dominant focus of the image is its portrayal of poverty. The protrusion of the horse's ribs seizes the viewer's attention. The horse's head is drawn with hunger. Lucy is in a pallid dress, but we don't see her face—only her profile, her hat pulled down to her eyes. She could well be doing a successful job hitching up the team, but the landscape surrounding her reinforces the impression of isolation and destitution. This image was more effective at catalyzing support for child labor reform than the images of Jack were—empty fields and empty bellies call forth greater empathy than Jack's work elicits. Hine was also conscious of his gendering of these images. From the perspective of reformers seeking signs of exploitation, did the circumstances of girls working in agriculture warrant different concerns from those of boys doing the same or similar work? By gender alone, did images such as Lucy's generate additional anxiety?

In other photographs, Hine made his critique more pointed and better aligned it with concerns over children performing farm labor or any labor at all. Hine photographed eleven-year-old Carl Brown raking hay on the 160-acre farm his father owned in southern Vermont. In a photograph of Carl working alongside his father—his

FIGURE 6.5 "Lucy Saunders hitching the team to the horse rake." (Hine 3979)

father standing on the ground raking hay while Carl stands atop the wagon, seeming to spread the hay evenly—Hine gave a new twist to an old theme in the child labor reform circuit. Hine's caption notes that Carl is "overgrown" and "sluggish." While large numbers of children abandoned their education to work because they didn't care for school, Hine reports that Carl said, "I'd ruther go to school."[5] Does this suggest that agricultural labor held fewer attractions for children than did industrial labor, the monotony of farmwork making even school look good?

In some cases, Hine's photographs and captions worked together to portray the dangers that child farmworkers were subject to. Hine made several photographs of William Stewart's family in August 1915; together with his wife and two sons, Clinton and Wason, Stewart ran a 135-acre farm in South Stephentown, New York, just across the border from Massachusetts. One photograph shows twelve-year-old Clinton operating a mowing machine that had "cut off his hand" the month before (fig. 6.6). Hine reported that, four weeks after the accident, the young boy was doing what he could to help out with the farmwork, even though his maimed arm was still in a sling. Clinton's mother apprised Hine of the consequences of her son's injury: now that Clinton had lost his ability to farm, Mrs. Stewart bemoans, "we will have to educate him."[6] His economic value to his family as an agricultural laborer was much diminished; thus, as his mother suggests, education was imperative. Clinton's accident may well have inadvertently elevated his future earning potential and social standing.

FIGURE 6.6. "Rural Accident. Twelve-year old Clinton Stewart and his mowing machine." (Hine 3987)

FIGURE 6.7. "Work that Educates. Twelve-year old boy tending bees." (Hine 3996-A)

## JOHN SPARGO AS A MODEL

For children working on the farm at a tender age, parental oversight could make the difference between exploitation and education. In general Hine took a critical perspective when photographing children working on family farms, such as those of the Stewarts, the Browns, and the Saunderses. However, when he photographed his friend the socialist writer John Spargo working alongside his son, Hine held them up as a model for others to emulate.[7] Hine stopped in Bennington, Vermont, in August 1914 to visit Spargo and photographed him and his twelve-year-old son performing varied tasks—weeding, feeding the chickens, tending bees (fig. 6.7). Hine captioned each photograph as "work that educates." The work looks more like teaching than toil. Hine seems to capture a tenderness between his subjects, as the boy tends bees "under the direction of his father, John Spargo." Spargo is supervising like a parent or an educator, not like an employer or a boss.

In Hine's photographs of Spargo and his son, the boy is not laboring in the same way Jack is in Hine's images of the young Massachusetts farmworker. The photographs of Jack depict a boy operating machines and doing men's work. Spargo's son, though older than Jack, appears younger and less hardy, and he carries out his chores under close supervision and direction. The photographs of Spargo's son represent an idealized depiction of farm labor. He is not overworked, nor is he driving dangerous equipment. The labor he performs is safe and expected of a young boy or girl growing up on a farm. Hine may have hoped that the photographs of Spargo and his son would evoke Spargo's authority as an "expert" on the subject of child labor. Advocates of wholesome farm labor for boys and girls were promoting the merits of children toiling not as wage laborers but as farmworkers. Family farm labor seemed to buffer the child laborer from the capitalist market, but the transition from working on a family farm to wage-based agricultural labor drew the child right back into that orbit of exploitation.

When Hine photographed agricultural wage labor instead of children working the family farm, he found the material he and the National Child Labor Committee were looking for—photographs depicting exploitation and abuse. Hine surveyed two forms of commercial agricultural labor: cranberry picking in Massachusetts and the cultivation and picking of tobacco in the Connecticut River Valley. These types of work—and specifically the employment of children in dead-end agricultural jobs for low wages—were of the greatest concern to the NCLC. If Hine's photographs depicting Spargo's relationship with his son conveyed certain virtues of agricultural labor, they also offered a most pointed critique of commercial farming that made extensive use of children as cheap wage labor. The images of John Spargo and his son served as an alternative model to the exploitation of children in commercial contexts.

### CRANBERRY BOGS OF MASSACHUSETTS

Following his investigation of sardine cutters and canners in Maine in August 1911, Lewis Hine moved south at the invitation of the Massachusetts Child Labor Committee. (Hine's services were often lent to various local committees by the national office.) He came to Massachusetts to investigate cranberry workers picking in the bogs of Plymouth County in the southeastern section of the state, near Cape Cod. As he had in Maine, Hine conducted a visual survey, but this time he did not work alone: Richard K. Conant, secretary of the Massachusetts Child Labor Committee, accompanied Hine during his investigation into cranberry farming and served, according to Hine, as a "witness." Although it isn't clear why Hine needed a witness when he photographed children such as Merilda carrying cranberries across the Eldridge Bog (fig. 6.8), Conant did verify the integrity of Hine's photographs. Together with Conant and Owen R. Lovejoy, general secretary of the National Child Labor Committee, Hine created a report, "Child Labor on the Cranberry Bogs of Massachusetts," to supplement the eighty-one photographs Hine took of child laborers, their families, and their

FIGURE 6.8. "Merilda. Carrying cranberries." (Hine 2546)

living conditions.[8] Through the combination of the report and his photographs, Hine sought to generate empathy for the fate of child farmworkers.

The commercialization of cranberry farming began after the Civil War and continued into the twentieth century, transforming the family cranberry farm into a more "capitalized agriculture." Swamplands were cleared, drained, and made into productive cranberry bogs. In Carver, Massachusetts, for instance, where Hine inspected local bogs, the amount of acreage under cultivation increased dramatically, from 750 acres in 1890 to 2,461 acres by 1912.[9] According to a Massachusetts census of agriculture, in 1905 the state had 1,939 independent bogs, with almost 70 percent of them, or 1,345 bogs, located in Barnstable County, and 26 percent, or 510 bogs, in Plymouth County. These two counties were responsible for an overwhelming share—nearly 91 percent—of the cranberries harvested in the state. The size of the bogs varied greatly, as did their level of capitalization, their productive capacity, and the composition of their workforce. As for marketing the harvest, almost all of the bogs, irrespective of their size, relied on two cooperatively run companies for the sale of their cranberry crops.[10]

There were many similarities between the circumstances of the sardine canners in Maine and those of the cranberry pickers in Massachusetts. Sardine canning and cranberry picking were both seasonal industries, though the cranberry picking season, which began in late August and continued into October, was much shorter than the sardine season. Also, sardine canning, like cranberry picking, was family-based labor in many instances. Unlike families working in the Maine canneries, however, a family working in the cranberry bogs more often labored together as a unit that was managed within the family and overseen by the boss. This tended to diminish much of the autonomy the children might have enjoyed, including receiving their own earnings. Another important difference between the two types of workers was that many cranberry pickers had other jobs to go back to at season's end. Some canners tried to make a year's income from the sardine season. Because of the shorter picking season, that seems to have been the case less often for the cranberry workers. For many of those who labored in the cranberry bogs, picking was an annual interlude that gave them an opportunity to leave behind the factory and the muggy city heat and work and sleep under the clear skies of Cape Cod for a few months.

The ethnic composition of the cranberry workforce was perpetually changing as new arrivals took the place of old hands, and native-born workers were replaced by Polish pickers, Italian pickers, Finnish pickers, and others. Each bog was different with respect to its employees' ethnic makeup. At the Maple Park Bog in East Wareham, where Hine photographed six-year-old Joe, his eight-year-old sister, and their mother taking a moment's respite from their labors, the majority of the pickers were Syrians who had come from Boston and Providence to work on the bog.[11] In South Carver, Hine photographed Alphonse Londry, a seven-year-old French Canadian boy, and his family working on the T. B. Smart Bog (fig. 6.9). The Londrys came from Fall River to pick cranberries.

Hine's report to the NCLC gave the impression that, with the ongoing influx of immigrants into the country, it was easy to replace one group of cranberry workers with another if employers encountered difficulties. For example, one bog boss told Hine that a few years earlier, when he had "had trouble with the Portuguese," he merely "went to New York. Got 40 Finns who made find [*sic*] hands." In any case, there was "very little native talent." According to Hine, this refuted the popular notion that pickers in the industry were "Cape Cod Yankees . . . out picnic fashion with their families."[12] Immigrant families, many with rural backgrounds, moved out of the textile cities to pick cranberries together and sleep under the stars each night.

The most significant development in the cranberry labor force was the widespread presence of Portuguese immigrants from the Cape Verde Islands. In the decade after the turn of the century, these immigrants had come to predominate in the cranberry workforce, with hundreds of Cape Verdeans hired over the course of the year and another three thousand or so at harvest time. By the time of Hine's survey, they were "proving themselves the best pickers and the best wheelbarrow men who ever came

FIGURE 6.9. "Alphonse Londry. Said 7 years old." (Hine 2520)

upon the bogs of Cape Cod." Each August, Portuguese came from the mills of New Bedford, from the docks in and about Providence and Fall River, from the ranks of oystermen and longshoremen, "here and there from out of the woods and wilds in the vicinity of the cranberry district, and by twos and threes, by gangs, and by hundreds," to work on the "fruit-laden" bogs of Plymouth, Barnstable, and Nantucket.[13] The vast majority of the Cape Verdean cranberry workers were peasants from the islands of Fogo and Brava, where small-scale farming formed the basis of the economy. Most had probably never seen a cranberry before their arrival on the Cape, but the similarities between the agricultural systems of their island homelands and the cranberry culture of Cape Cod aided them in making the transition to the cranberry economy.[14]

The "cranberry district" was the only place in New England in which Hine photographed a largely nonwhite workforce (fig. 6.10). Hine reported that "in all of the foregoing bogs we found not one where the natives were used to any extent." Most of the cranberry workers he surveyed were "Portuguese," and "throughout the district" he found "the almost universal use of Portuguese [was] noticeable." The Cape Verdean immigrants, though classified as Portuguese, were of dark and mixed complexion, with strong African ancestry. Notions of race were remarkably fluid in this period of

FIGURE 6.10. "Even the little tiny one about 5 years old was picking." (Hine 2536)

immigration, as Matthew Frye Jacobson illustrates in his book *Barbarian Virtues*.[15] Racial identity was particularly complex for Cape Verdeans, whose self-identification as Portuguese speakers was often at odds with the perceptions of those around them in New England.[16] Hine touched on this complexity in his report, noting that these workers, "as black as negroes," were "really Portuguese called Bravas."[17]

Of course, being black in the United States in the first decade of the twentieth century meant having to withstand significant intolerance and ignorance. In many parts of the country, "lynch law" ruled the day, bolstering a new plantation system based on sharecropping, black disenfranchisement, and white supremacy.[18] In New England, the determination of many to view the Bravas as blacks left them open to racial stereotyping and racist condemnation. Some of these attitudes can be detected in a 1911 profile by the Dillingham Commission on Immigration, which gave the Bravas a mixed evaluation. The commission paid them a backhanded compliment, declaring that "the Bravas have a much higher moral code than one would suppose, judging by their ignorance and their standard of life, . . . and are generally temperate." The Bravas engaged in little crime, minimal gambling, and modest drinking, according to the commission. Nobody contested their hard work and overall frugality. However, although the commission found that almost all of the Portuguese, "white as well as black," were illiterate, it stressed that the Bravas in particular were "stupid," had "little regard for a Puritan Sabbath," and showed almost no concern for cleanliness or sanitation in their homes: "everything dirty . . . is typical of a 'Brava' household."[19]

The local schools did not segregate the Brava children, but the Dillingham Commission noted that problems were beginning to arise in cranberry communities as the Bravas developed "a growing sense of their importance." Electric car conductors, trainmen, and others complained that "some of the Portuguese are beginning to have 'rights.'" The report seemed to lament that "a few years ago it was easily possible to put a Brava on the back seat in a street car and to make the 'jim-crow' car idea a practicable expedient." Now, it bemoaned, the "Brava who knows his importance refuses to move back or forward or anywhere else until he pleases to do so, much to the annoyance of the conductor." An "old electric trainman" complained that things were getting worse, to the point that some of the "bolder spirits seem to take delight in sitting down in the same seat with a white woman, if there is any opportunity. The white patrons complain, but there is no legal method of putting the Brava into any seat he does not choose to occupy."[20] Old New England felt threatened by assertive black men and women who refused to accept the color line and who failed to see any reason they should defer to whites.

Hine did not show any evidence of sharing those sentiments. Although he did not express a sophisticated understanding of the Cape Verdeans, he nonetheless accorded them the same respect he showed white workers. Hine photographed Brava children in sympathetic portraits, Brava families as intact units, and Brava men and women as human beings. In the history of American visual culture, there seems to have been almost a necessity to demean people of color: in movies, plays, cartoons, and editorial illustrations, people of color were invariably cast in a negative and dehumanizing way. Hine felt no need to emasculate his subjects in his photographs, or to fit them into existing caricatures. He did not photograph to tear down or demean but to portray the dignity of his working subjects, young and old, black and white alike.[21]

We of course see only the people and places Hine successfully photographed, but Hine's field report is overloaded with descriptions of efforts to photograph child laborers that failed, either because the bogs were closed or because they did not hire children. For instance, on August 22, Hine went to the Sparrow Mills Bog near Marion, but the workers were not picking that day. Hine and Conant visited a "private bog" near Parker Mills on the 13th of September but found no children working there. We don't know why Hine didn't photograph at the Hollow Brook Bog, where he reported that seven of the fifty-seven workers, according to their own testimony, were between six and twelve years old and worked from nine in the morning until five at night. At the Tweede Barnes Bog, he saw no children working. Likewise, at the Robins

Bog, the East Head Bog, and the four bogs of the Federal Company, he found only adults employed. At the Frog Foot Bog and the Five Acres Bog, both owned by the Makepeace Company, Hine determined that only two out of fifty-two workers were younger than twelve years of age. Hine and Conant reported that eight of the fifty-five workers employed at the Smart Bog near South Carver, Massachusetts, were between the ages of six and thirteen. Hine visited a dozen additional bogs with Conant, again with very mixed results. No one was working on the Bangs Bog and the Wankinoo Bog on the days they visited, and a small bog beyond Five Acres Bog employed no children that Hine could see. The irregularity of the industry's employment practices was clearly evident, but so too was the fact that child labor was not universally present in the bogs of Plymouth County.[22]

While the cranberry industry appeared profitable, it was the owners and stock-holders, not the workers, who reaped the lion's share of the benefits. A manager at Swift's Bog near Falmouth said to Hine, "We have 150 workers, besides the kids. . . . There is good money in the cranberry business." A bog, Hine reported, could yield between seventy-five and one hundred barrels per acre. A barrel that generally sold for $14 could usually be produced for only $1, netting the bog owners considerable profit on their investment, although bosses at Baker Bog put the net profit per barrel closer to $5.75. As for the stockholders, the Swift's Bog manager told Hine that they "are never satisfied until they make 25 to 35% dividends."[23] Meanwhile, at eight cents for a six-quart pail, no pickers got rich. Adult pickers earned approximately $80 in a very good year, but their work during the 1915 season, which had not been a particularly good one, might pay them only $50 in total, according to Hine. A picker who was discharged for any reason received "only 6 cents for what he [had] picked." Discharges, Hine noted, were frequent, especially "when the bog is crowded because of picking odds and ends."[24] Child pickers labored for pennies.

As was true of many agricultural and extraction industries, profits in the cranberry industry derived in part from the bosses' monopoly on weights and standards of measurements. A pail held six quarts, and a barrel was supposed to hold thirteen and one-half pails. However, one of the bosses showed Owen Lovejoy how they "packed the barrels—filling them very full then putting on two or three more quarts of berries and pressing the barrel head on." Hine noted that the "checkers" who received the barrels got into "the habit of saying, 'You must fill your measures fuller,'—though they were most of them heaped anyway."[25] It is not clear whether children like Mary Christmas who picked up berries that had fallen to the ground around the barrels were paid for their labor or were able to put together their own pails from the spilled berries.[26]

The method of picking affected the measurements as well, because different methods generated different amounts of waste, and the level of waste in a pail affected the picker's pay. Berries were put through screening machines to remove the waste, which was deducted from the picker's haul, thus reducing his or her pay. If there were vines or stones in the six-quart pails, pickers were "paid for only 5 quarts." Handpicking the cranberries generated the least amount of waste product, while picking them with scoops generated greater amounts of waste. One boss showed Lovejoy small "boxes of waste under the table from the hand pickers of about 16 quarts representing three days work." At Week's Bog, where scoop picking predominated, Hine saw significantly larger amounts of waste.[27]

Using scoops not only was "faster" but also cut down on the number of children working on the bogs. Hine noted that the Manomet Bog had previously hired women and children as pickers, but, "now they use scoopers entirely and that is work for men." In fact, other than the Eldridge Bog, which Hine stated "was behind the times," all the bogs used scoopers. Although scoopers were men most of the time, the predominance of male pickers raised the cost of labor. One boss got around this by making some of the scoops "smaller for the children's use."[28] Hine photographed Charlie Fernande showing off the smaller scoop he worked with (fig. 6.11), whereas his photograph of Carrie Maderyos shows her carrying a larger scoop that she never would have been able to operate at her age and size.[29] Hine also photographed ten-year-old

FIGURE 6.11 "Charlie Fernande showing the scoop with which he works." (Hine 2513)

FIGURE 6.12. "Gordon Peter, using scoop with metal teeth not covered." (Hine 2518)

Gordon Peter (fig. 6.12), who had picked for three years and had graduated to a small scoop with metal teeth.

Bosses often further increased profits by "sweating" their workers. If pickers didn't meet expectations, neglected to pack their pails tight with cranberries, or failed to clean their pails and barrels of vines and other matter aside from berries, they suffered abuse and could expect lighter envelopes come payday (fig. 6.13). Hine reported on one boss, an old sea captain, who "threatened" his workers that they needed to hit the mark or else they'd "go ashore." The boss at the Hollow Brook Bog was particularly harsh, "a veritable slave driver." Hine, whose experience with different forms of work was considerable, was reminded "of the speeding up system in a sweat shop. They work on the run pursued by all the epithets the boss could fire at them"—much of it, Hine added, was "profane." When a "towsly headed boy about 10 years old" came to the checker with a box of cranberries delicately and precariously "balanced on his head" and "struggling to keep it up" (fig. 6.14), the checker, who apparently was mild next to the field boss, shouted at the boy: "Put it in there, god damn it. Hold on, god damn it, go back and fill it up." When told that there was no need to curse, the checker merely replied, "Well, I ain't cursing, god damn it, but go back to and fill it up." When another boy headed his way from across the field, similarly balancing a

FIGURE 6.13. "Group of Workers on Smart's Bog." (Hine 2521)

FIGURE 6.14. "Boy who carries barrels. Robert Saunders, 10 years old." (Hine 2524)

box, the checker shouted, "Go back and fill it up or god damn it you'll go home." On another occasion, Hine reported on a little boy of twelve, picking vines from one of the barrels, who was accosted by a checker yelling at him to "Take them vines out. Throw em to hell overboard." The checker continued with a "running fire of this 'speeding-up' sweatshop talk."[30] The torrent of abuse heaped upon the young workers was unacceptable to Hine, who viewed childhood as a precious time.

Since picking cranberries was seasonal, many of the cranberry workers were migratory. Most of them, according to Hine, lived "around the neighborhood" or came "from New Bedford." Many also moved out of Fall River to join the cranberry workforce each harvest. Hine found housing conditions for these workers "usually fairly good," but "there were [sic] some crowding at the Douglas Farm, Maple Park Bog, and a few others." Mrs. Phinney, who owned one of the bogs that Hine and Conant visited, told Hine that she knew of a bog for which twenty-one workers lived "in one small house," where she thought "they must sleep standing up." Overall conditions mattered little, according to one bog boss. He made trips to Fall River and New Bedford each summer to recruit families. He felt that "his workers are in the factories the rest of the year and give up their jobs for a time to come down here and get a holiday." Just to ensure that workers stayed for their entire "holiday," they were not paid until the end of the season, "or they would all leave at any time."[31] Withholding wages was a powerful tool to ensure the workforce was there when needed.

Hine did not photograph workers' migration from the textile factories to the cranberry bogs, but he did create a couple of images that generate a sense of movement, a notion of a family "going to work." In one image Hine's subjects are walking away from Hine, while in another they are looking directly at the camera (fig. 6.15).[32] They are clearly recognizable as "foreign" and are just as clearly recognizable as a family, though they are certainly not that Yankee family going for a picnic; rather, they represent Hine's symbolic call for hard-working immigrant families to be Americanized and brought into the new nation as citizens.

According to the U.S. Department of Agriculture, cranberry growers in Massachusetts generally provided their employees with little in the way of housing or other amenities, such as food or cookware. Most growers had "no accommodation for their workers except that most of them have shacks where they may prepare meals and live and sleep in mild weather." Pickers generally spent only a few weeks in early fall in the bogs. Large number of pickers who worked in the industry, particularly the Bravas, had homes in cities and towns not more than thirty miles away. Many came from Providence, Fall River, and New Bedford, where they labored in "city work and industrial employment." Some pickers lived farther away, including a few who migrated as far south as Virginia to work the "truck crop harvests" before returning north in the fall.[33] For many workers, cranberry labor was a nice complement to factory work.

FIGURE 6.15. "Going to work." (Hine 2557)

As was the case for school-age workers in other seasonal industries, education was a problem for child laborers among the cranberry workforce, who frequently didn't start attending school until October. Amelia, a twelve-year-old cranberry picker in the third grade, which was then in session, told Hine that she had received a visit from the school superintendent (fig. 6.16). Amelia, who earned seven cents a pail for the twenty-two pails she picked each day, said that the superintendent "came to our Bog today and said we got one more week to pick." Parents no doubt had mixed feelings about sending their children back to school. A "Portuguese father" working near Week's Bog said that he could make as much as six dollars a day when he had "several others picking in his family," and he was not about to give up that precious income. He ignored the superintendent, telling Hine that "lots of children pick after school begins." A seventy-eight-year-old Portuguese man bragged to Hine that he had fathered thirty-eight children and that all of the surviving children—thirty-four in all, ranging in age from two to fifty-six years old—worked in the bogs, making his family's labors very profitable indeed. Picking cranberries was a family affair: men and women, the young and the old, adults and children all labored side by side in the bogs.[34]

FIGURE 6.16. "Amelia Louise Sousa, 12 years old. Picks 22 measures." (Hine 2576)

## TOBACCO IN CONNECTICUT

In August 1917, as Hine was nearing the end of his work with the National Child Labor Committee, and prior to his departure for Europe to photograph for the Red Cross, he journeyed to Connecticut to survey and photograph the work of children in the tobacco industry. Hine traveled to the northern part of the state, the "valley of the Connecticut River," where he moved among "hundreds of tobacco farms ranging in size from a few to several hundred acres." Thousands of people toiled in the cultivation and harvesting of Connecticut-grown tobacco. Although the industry was well established in the state, Hine believed that few people, "even those in closest touch with the industry, . . . realized to what extent children, are involved in this work." In the course of his investigation that summer and early fall, Hine witnessed "several thousand children from eight to fifteen years of age daily employed in the tobacco fields and sheds." The extent of their participation was a "shock" to Hine; many managers in the industry had told him that the number of child workers was "relatively small and constantly decreasing," but the evidence from his investigation told a different story.[35]

Hine's photographs of field-workers for the American Sumatra Tobacco Company in South Windsor, Connecticut, demolished in a glance any suggestion that few children labored in the tobacco fields (fig. 6.17). After war broke out in Europe in 1914, munitions factories in cities such as Bridgeport, New Haven, and Hartford were swamped with orders for small arms and ammunition. In the tobacco fields, children became the workers of choice because they often were the only ones available for the job. More important, their labor was inexpensive relative to the cost of adult labor. The enforcement of child labor laws was modestly relaxed, and school and work calendars were remade to accommodate each other. Thus, when the superintendent at Sumatra Tobacco called his field-workers to the end of the tobacco rows, Hine unsurprisingly counted forty-seven boys between nine and fifteen years old. The superintendent said plainly, "We have to hire boys because we can't get men to do the work." Hine recognized the aggressive efforts of the Connecticut Labor Bureau to "get workers over sixteen years of age," but during harvest time, when labor needs were particularly acute, employers maintained that their inability to secure adult workers "compelled" them to hire children. A few tobacco growers said that they employed "practically no children under sixteen years" of age, but Hine later learned that the particular kind of tobacco they grew called for "older workers."[36]

Despite the abundance of child workers, tobacco growers in Connecticut saw little need to reform their labor practices, as many of them believed and publicly insisted

FIGURE 6.17. "Field-workers, Amer[ican] Sumatra Tobacco Co." (Hine 4875)

that labor in the tobacco fields and sheds was beneficial to children. Hine spoke with many owners and managers who rationalized the practice of employing children by claiming that it helped to instill in children "the habits of industry, etc." The principal argument from the growers' perspective was that working conditions for children were not as harsh or dangerous in the tobacco industry as in some other fields of industrial employment. Hine did not dispute their point; he merely expressed his concern that the tobacco industry presented enormous possibilities for the exploitation of children.[37]

Hine did, however, speak with managers and assistants who mistakenly believed that current conditions of child labor were already protected and regulated by state child labor laws dating back to the 1840s. From the outset of state and federal child labor legislation, agricultural workers had been left unprotected, a testament to the powerful interests of commercial farming and the romantic ideals derived from common notions of the family farm. Hine stated that some of those in charge at Connecticut's tobacco farms "were under the impression that the state labor law applied to this kind of work and one told me that he was enforcing an 8-hour day for children under sixteen because he did not believe that they should work longer hours." Such employers, however, made no reference to the great struggles in the state—emanating from urban industrial centers, most notably Bridgeport—to establish a general eight-hour workday. Efforts to make any employee work more than eight hours in a day undoubtedly worsened labor relations, intensified labor conditions, and contributed to the labor shortages in the tobacco industry. According to Hine, when one particular employer learned that tobacco workers were not covered by state labor laws, "he contemplated going back to a ten-hour day for all."[38] Because they were outside the protection of the law, children working in the tobacco industry were clearly vulnerable.

Hine began his investigation into Connecticut's tobacco farms in early August, visiting more than fifty farms, among which were the largest tobacco growers and corporations in the state. At each site Hine made contact with the owner, superintendent, or foreman, and under their direction he counted the number of children at work in the farm's sheds and fields. He also determined each child worker's age by questioning either the child or the child's parents. In all, Hine found 1,458 children from eight to fifteen years old working in the sheds and fields; of this number, two-thirds were between the ages of eight and thirteen. Hine believed that these numbers should actually be doubled for a more accurate accounting: he had "covered" only 60 percent of the "shade-grown" tobacco fields and only a very small number of the "sun-grown" fields, many of which were smaller plots employing only a few adults and children. Although the sun-grown fields were important, their size and dispersion throughout the region required "a great deal of time to visit"; thus Hine left them out of his survey.[39]

As it did in other industries, the number of children employed in the tobacco industry of Connecticut varied from day to day. Hine generated a detailed report, "Record of Ages and Children Found Working in Sheds and Fields of Connecticut Tobacco Farms," in which he endeavored to portray "fairly typical conditions through the season." Based on his findings, the use of child labor on tobacco farms was extensive. Hine noted that at thirteen sheds he surveyed, 30 to 50 percent of the child workers were under the age of fifteen. Seven sheds had even higher concentrations of child workers who were fourteen or younger—between 50 and 60 percent—while from 60 to 70 percent of the children working at six other sheds were under fifteen years of age. Hine found four sheds in which children under fifteen years old made up 70 to 80 percent of the child workers, and another shed in which 87 percent of the child workers were under fifteen; in still another shed, all of the workers were between eleven and fourteen years old. Within these rates of concentration of child workers, Hine reported that the number of children as a percentage of the total workforce was higher in the fields than in the sheds.[40]

Hine found that the child labor force in Connecticut's tobacco industry was clearly divided by gender. In southern tobacco fields, boys and girls usually performed the same work, with the exception of a few tasks requiring "a great deal of physical strength," but in New England, particularly during harvest time, boys worked outside in the fields and girls worked inside the sheds. Not only did boys' and girls' work differ, but girls were often paid under a system of piece rates in the sheds, while boys working in the fields received daily wages. Pay varied for both boys and girls according to their age, with the younger and less experienced children earning less. Some boys also worked in the sheds "for pickers, but often did one or more of these tasks on the same day they did some picking." Hine also noted differences in the racial composition of the child workforce on northern and southern tobacco farms: in the South, African American children composed one-quarter to one-third of the total workforce, whereas black children laboring in the tobacco fields of Connecticut was a far rarer sight until the First World War began and increased the black migration northward. In the North, native-born children of immigrant parents predominated in the Connecticut tobacco fields.[41]

Hine photographed the gender divisions of labor in the tobacco industry, capturing the boys who picked the shade tobacco, for example, while also attempting to give perspective to the vast fields of tobacco. In several photographs he took of a "second picking" on the Goodrich Farm in Cromwell, Connecticut, Hine has the viewer look into the horizon. In one of those images, a diagonal line made of boys standing at the end of rows disappears into the horizon.[42] For another image, Hine found a way to take the photograph from an elevated level, creating more of a bird's eye view of the expansive fields that seem to go on forever (fig. 6.18). Both images give a quick read of the extensive use of child laborers. The images are essentially the same scene but are

FIGURE 6.18. "Tobacco pickers on Goodrich Farm, 'Second picking.'" (Hine 4878)

crafted from slightly different angles to create two distinct perspectives. Both depict the vast fields of tobacco punctuated by the light-colored shirts of dozens of boys.

Hine photographed both boys and girls working in the sheds. One such photograph shows six "leaf-girls" in Weatogue posed in a group portrait, with each girl holding a pile of tobacco leaves (fig. 6.19). The girls would flatten and prepare the leaves before handing them on to the next group of girls, who strung the leaves and then passed them along to be hung up in the drying shed by "leaf-boys," such as the three boys, two of them nine years old and one eleven years old, whom Hine photographed at the Cybalski Tobacco Farm in Hazardville.[43] A group shot of shed workers at the Wetstone Farm in Vernon, Connecticut, shows both boys and girls, including some slightly older boys sitting up on a rafter above the group.[44] These older boys would scamper across informal scaffolding, casually constructed to reach all the rafters throughout the shed, and hang as much tobacco as possible. Of all the tobacco occupations, this could be the most dangerous, not only because of the rafters and stringers the boys had to walk but also because of the threat to the boys' health from the heat in the top of the sheds and the intensity of the drying tobacco, which burned their eyes and often nauseated them.

Work in the tobacco industry was seasonal and paid unusually high wages. Managers reported to Hine that the season was a short one, generally calling for only thirty to fifty

FIGURE 6.19. "Leaf-girls (4896) 3 of 11 yrs., 2 of 12 yrs., 2 of 13." (Hine 4897)

days of actual work, which usually began in early August and ended sometime around mid-September. Hine had expected his survey to show that wages were high for this kind of work, but he found the earnings of the better workers "astonishing." According to Hine, earnings varied by age. The youngest workers were eight years old and earned between 60 cents and $1 a day, while nine-year-old workers generally earned between $1 and $1.25, with a few taking home $1.50 a day. Daily wages for ten-year-old workers were $1.20 to $1.50, though sometimes more, and eleven-year-olds' wages rose to $1.50 and up. Twelve-year-olds received anywhere from $1.50 to $2 per day; thirteen-year-olds were paid $1.60 to $2 for a day's work; fourteen-year-olds earned anywhere from $2 to $2.40, with one beginner starting at $1.60 per day; and fifteen-year-olds could earn between $2 and $2.80 a day.[45] By comparison, a messengers could earn only $5.00 a week. It is little wonder that children and their parents sought employment in tobacco.

Hine believed that the high wages for child tobacco workers would serve as an incentive for employers to examine the labor process and seek out opportunities for mechanization, which would reduce the demand for child laborers. Specifically, Hine recommended a "need for better organization of the labor force and improved methods of handling the leaves." He believed that the methods for stringing leaves on laths with a needle and hanging the laths in the shed to dry were "primitive." Hine maintained that a simple device that could string and hang the leaves would "greatly reduce the need for children and facilitate the handling of the leaves when they are ripe."[46] Child labor had a tendency to impede mechanization, however; as long as labor was inexpensive enough—and child labor was cheap labor—it was unlikely that producers would invest too heavily in machinery. And when the supply of one group of children moved on, industry representatives sought out new labor elsewhere.

Hine commended the efforts of the Connecticut Leaf Tobacco Association, which he believed was "doing good work organizing the distribution of labor in these fields." Its greatest achievement, Hine believed, was in recruiting labor from outside the region. The association had been able to bring approximately a thousand adult workers from New York City to Connecticut. It also did an admirable job, according to Hine, of recruiting "colored students" from southern colleges and encouraging "colored workers" from elsewhere to seek employment in the fields.[47]

The Connecticut Leaf Tobacco Association also recruited labor from within the Connecticut Valley region. "Hundreds of women and children" were transported from Hartford and other Connecticut cities by "auto trucks and trolley cars" every morning and then were returned home at night. According to Hine, the association "constantly" emphasized the need for workers over the age of sixteen, but when the demand for labor was urgent, as it invariably was during harvest season, the growers were "inclined to accept workers of any age provided they can do the work." As the war in Europe continued and the labor shortage became more acute, the "demands for young workers . . . further intensified."[48] Hine photographed the open-air trucks that carried workers to the fields from Hartford and other cities. One of these photos shows a large group of workers, many of them young girls, boarding a truck in Post Office Square in Hartford at six in the morning (fig. 6.20). Children often had significant travel on either side of their employment, but this transport service made it possible for large numbers of urban children to work in the countryside.

Hine found the work conditions for children who labored in the tobacco fields to be good for the most part. The heavy work was done primarily by adults, with "notable exceptions such as the cases of young boys dragging heavy baskets and climbing aloft in the sheds hanging the leaves on the rafters." In general, Hine found a "good wholesome atmosphere prevailing everywhere." He was impressed by the "high grade" superintendents and foremen he encountered and the "real consideration" they showed the workers. Many of the young workers were there with family members, and among these groups Hine noted that "a most delightful spirit of industry existed." The sheds were cool and pleasant even on hot days, in contrast to conditions in the fields, where exposure to the heat of the summer sun "seemed more serious especially when the children were working on the ground picking the lower leaves and the tall plants kept off the circulation of air." Young workers in the fields were also more removed

FIGURE 6.20. "6:00 A.M. at Post Office Square. Truck load of tobacco workers." (Hine 4893)

from supervision. Hine commented on the potential for harm to befall young women under such conditions, but in only "one case were young girls found working in the fields and this was reported back to the superintendent at once pointing out the moral hazard involved in such laxity."[49]

While the work was not hard and the conditions were not overly taxing, the length of time that child workers labored in the fields and sheds each day was more problematic for Hine. Almost all of the growers reported that the tobacco workers, young and old, performed "actual work" for nine to nine and a half hours per day. That amount of time, Hine argued, was "altogether too long for young girls to stand as they do stringing and handling the leaves in the sheds." Hine advocated an eight-hour day for child workers.

Hine also reported that health-related issues did not appear to be a significant concern for the young tobacco workers. Several organizations, most notably the Consumers' League of Connecticut, expressed satisfaction with the workers' overall safety. Few injuries seemed to have resulted from the work. The one threat to workers' health came from handling the tobacco leaves, and this tended to have only a temporary effect: new workers, according to Hine, often experienced some feeling of "nausea for a day or two when the work is begun." Other than this, and aside from his belief that additional supervision might "forestall any possibility of exploitation or over-work," Hine found the industry free of any "glaring disadvantages."

Hine was pleased that the matter of school attendance did not appear to be a serious problem for the young tobacco workers. He found "little or no interference with school attendance." The seasonal quality of the work coincided well with local school schedules, with school often beginning just as the tobacco season ended. In cities such as Windsor and East Hartford that provided a good share of the tobacco labor force, the school year often started a week or ten days later than it did in other cities to accommodate the working children, pushing the conclusion of the school year forward by an equal amount.[50]

Of all the industries that Hine surveyed, tobacco picking and processing seemed to him the least objectionable, or perhaps Hine himself was changing. This survey was among Hine's last assignments while he was employed by the National Child Labor Committee. Was he becoming bored with his work? Was he preoccupied by the war? Was he ready to move on to new things, such as his assignment for the Red Cross? Was he troubled by the NCLC's decision to cut his pay? All or none of these factors could have been at work—we simply don't know—but Hine raised no call for reform of the tobacco industry, other than appealing to commercial farms to mechanize some of the tasks done by children and thus eliminate the need for that cheap labor. In his survey, Hine found tobacco picking, with few exceptions, to present modest dangers. He noted a lack of worker exploitation in the industry, and he seemed to encounter, or report, fewer complaints from tobacco workers than he recorded during his surveys of other industries. He only briefly mentioned the very serious issue of workers' health. A full understanding of the threat posed to the health of children working in the tobacco industry would require the insight of new industrial health experts such as the lawyer and journalist Crystal Eastman and the industrial chemist Alice Hamilton, who were making significant contributions to the fields of occupational safety and toxicology. Hamilton's investigations of hazards in the workplace might have improved the lives of child tobacco workers and kept them away from the toxic product at the center of the industry.[51]

Throughout U.S. labor history, agricultural workers had continually been left outside the protective framework of labor laws in general, and of child labor laws in particular.[52] To be sure, working on a family farm was different from working on a commercial farm, and the challenges facing agricultural workers were different from those confronted by boys and girls working in the street trades and in textile mills. Agricultural conditions were not as pressing an issue as industrial conditions in the minds of reformers. Photographs of agricultural workers represented only about 10

FIGURE 6.21. "'Want any more MEN?' 7 year old Alec applying for job." (Hine 4903)

percent of the photographs Hine made of child laborers in New England, though the majority of the region's child workers, as elsewhere in the nation, labored in agriculture.

One of the final photographs Hine made of children on the Connecticut tobacco plantations, and one that was popular among reformers, was of Alec, a seven-year-old looking for work (fig. 6.21). The NCLC used the image in a couple of its publications and posters. Hine's photograph places Alec in the center of a dirt road that disappears into the horizon. The camera and photographer peer down on the boy, making him look even more diminutive in the picture frame. The boy wears overalls and a cap and looks humble, his hands tucked into his pockets, as he asks for work: "Want any more MEN?" He seems unaware of the irony in his request. Until the industry invested in machinery that could replace children, Alec's question was a valid one. With the labor shortage resulting from the war, Alec's worth might very well have bottomed out at a dollar a day, which was "the cheapest" he said he'd work for. Leaving aside any pressure brought to bear on the industry by mechanization and rising wages, Alec remained the labor of the future. After all, many of the young men who went off to war eventually returned. Tobacco owners must have been wondering, how are we going to keep them down on the farm after they've seen Paris?

## CHAPTER 7

# Exhibiting Child Welfare

———

A S WAS THE CASE WITH the "1915" Boston Exposition, early twentieth-century
reformers on both sides of the Atlantic seized upon the exhibition as a method
of communicating with the public and marshaling civic pride. Organizers of exhibits,
which took place annually at national conferences, sought to broaden discussion by
making public what were previously private conversations among specialists, their
professional organizations, and their volunteer affiliates. Their ultimate goal was to
bring together not only the reformers but those to be reformed as well, using photographs and other graphic materials to bridge social and cultural divides. City schoolchildren often assisted in this effort by serving as conduits to the new culture for their
immigrant parents. Exhibits were one place where the social met the scientific in the
emerging social sciences. The energy generated from these civic events was reckoned
by some to be like a great democratic revival washing over the landscape.

Child welfare exhibits were the most popular type of exhibit among Progressive
reformers. Children, whether real or idealized, possessed a unique power to create
a common cause among disparate groups. As flesh-and-blood sons and daughters,
they connected families to the idea of child welfare, and as an ideal they possessed
a civic value as harbingers of the national future. No other issue, save war, could
unite a city's people like children's welfare could. Child welfare exhibitions brought
together numerous reform campaigns and campaigners to stage collective portraits of
the economic and social ills facing communities. Professionals applied their discrete
knowledge to reveal the character of these community-wide problems and to offer
promises for human advancement based upon the latest developments in education,

social work, public health administration, government policies, housing, and urban planning. Child welfare exhibits represented an attempt to educate members of the public by appealing to their shared investment in their children, even if class distinctions meant that those who viewed the exhibits did not all sacralize childhood in the same way. Again, Lewis W. Hine's social photography was ideally suited to such efforts.

Child welfare exhibits sprang up in both large and small cities throughout Europe and the United States. Berlin, London, Dublin, Montreal, Chicago, New York, Philadelphia, Kansas City, St. Louis, Buffalo, Louisville, Knoxville, Rochester, New Britain, Northampton, Peoria, Toledo, Seattle, Indianapolis, and Providence—all staged industrial or child welfare exhibits. In March 1907, a committee led by Ellen M. Henrotin, a prominent reformer and a former president of the Chicago Woman's Club, put together the enormously successful Chicago Industrial Exhibit. In her foreword to the *Hand Book of the Chicago Industrial Exhibit,* Henrotin noted that "exhibits . . . of the sweated industries" had recently been mounted in Berlin, London, and Philadelphia, "and so great was the interest shown by the public that the committee felt it the psychological moment for a similar exhibit in Chicago."[1] Visiting nurse Currie D. Breckinridge reviewed the Chicago exhibit for the *American Journal of Nursing,* marveling that "with only a few weeks for preparation, the people of Chicago united to produce the Child Welfare Exhibit, an undertaking unsurpassed even in Chicago by anything except the great World's Fair of '93."[2] *The Encyclopedia of Sunday Schools and Religious Education* (1915) reported that more than one and a half million people had attended child welfare exhibits in recent years, and "in many cities the child welfare exhibit holds the record attendance among large gatherings and exhibitions." Although the exhibits *concerned* children, they were not made *for* children; yet children who viewed the exhibits apparently were "as deeply impressed as the adults." A recurring sentiment among social workers, businessmen, working men, and mothers was that child welfare exhibits were "the biggest thing that ever happened in our city; this will bring together and strengthen all work for the children."[3] Cooperation was the watchword in urban exhibitions.

Very few projects generated the kind of enthusiasm that welcomed the child welfare exhibits. In New England there were large exhibitions, such as the Child Welfare Conference and Exhibit in Providence, Rhode Island, and less expansive but still very effective exhibits, such as those in Northampton, Massachusetts; New Britain, Connecticut; and elsewhere. Northampton's was the first "small town" exhibit, and its success spawned dozens more. The demand for experienced exhibit directors and advisory literature was enormous, leading to the establishment in early 1913 of a National Child Welfare Exhibition Committee, which sought to provide assistance to communities seeking to stage child welfare exhibits. The committee offered a variety of services with the goal of propagating the ideas and research results of the Children's Bureau. It supplied communities with information regarding past exhibits, helped organize the committees necessary for mounting such expositions, assisted in planning the events, and when necessary, recommended and even trained exhibit directors. Shortly after the national committee was formed, Massachusetts organized its own committee "to promote and direct exhibits in the cities, towns, and rural communities of that state. This was largely the result of the Northampton success."[4] The popular exhibit form generated civic excitement and a cooperative spirit.

Anna Louise Strong, director of exhibits for the National Child Welfare Exhibition Committee, reported at the fortieth annual National Conference of Charities and Correction in 1913 that the term "child welfare exhibit" had become so popular that it had been "applied and misapplied to exhibitions of widely varying character, some semi-commercial, some philanthropic and educational, but limited in scope." For example, an exhibit prepared by a visiting nurses association on the proper care of infants was not a child welfare exhibit, in Strong's opinion, though it might be a "very excellent baby-saving show."[5] According to Strong, the term "child welfare exhibit" should be applied only to exhibits that aimed "to bring together *all forces in the community* dealing with the welfare of the child." In these exhibits, child welfare specialists organized themselves to present information on the city's children in a didactic manner, intending to enlist public participation in a program of civic exchange and social uplift by showing how local conditions affected children, what was or was not being done for children, and what ought to be done for them. Such all-inclusive exhibits differed from specialized exhibits on topics such as tuberculosis or housing in that the former required the "cooperation of all the community's forces." According to Strong, "securing of such cooperation is as much a part of the work of the exhibit as is the actual display." In many ways, these exhibits had the effect of a "civic revival": they called together all the social forces of the community to work out a plan for community action and then exhibited that plan "to citizens of all types and classes."[6]

Child welfare exhibits certainly benefited the participating child-helping agencies—children's institutions, settlements, boys' and girls' clubs, and philanthropic agencies dealing with children, for example. Financial assistance was the least of their gains; of greater importance was cooperation between organizations so that they might collectively reach beyond the limits of any single organization or campaign. In her report to the charities and correction conference, Strong relayed the sentiments of a Kansas City social worker whose organization had recently collaborated with other agencies in assembling a child welfare exhibit: "The exhibit shows us at one and the same time, the importance and the inadequacy of our work, and, above all, the tremendous importance of supplementing the best we are able to do, by the closest kind of co-operation with all allied forces." Civic altruism was at the core of the movement, Strong argued, and "any organization that is worthy of the name of *social* agency goes into an exhibit and into other work as well, not merely for what it can get for itself, but for the sake of serving the community."[7] The child welfare exhibit encouraged individuals, organizations, and institutions that shared a spirit of civic betterment to ally with each other in a cooperative effort to display their particular concerns and expertise as a collective whole for the education of the public.

Child welfare exhibits primarily affected a community in three ways: by spurring new laws, institutions, or officials; by educating people with certain facts; and by diffusing a general public sentiment in favor of taking action on behalf of children's welfare. The opening of the Rhode Island Child Welfare Conference and Exhibit in Providence in January 1913 under the direction of Anna Louise Strong coincided with the start of the state legislative session. The exhibit had a powerful, "almost humorous" effect on the Rhode Island legislature, recalled Strong, as lawmakers fell over one another to climb "on the band-wagon" promoting child welfare. Legislators introduced nine bills relating "to hours of labor, newsboys, a juvenile code and court, wider use of schools, the care of deaf, blind and imbecile children and several other questions." Publicity was the grease that moved reform through legislatures, counseled Strong: "A bill for increased appropriation to the State Home and Aid Society was an immediate result of the Providence exhibit."[8] Exhibits awakened the public and politicians to social issues and fostered legislative action.

In many instances, the effects of child welfare exhibitions were both immediate and lasting. Strong reported speaking with a writer for the *Kansas City Star* two years after the Kansas City exhibit. He told her, "I can get a hearing for any program of social reform by beginning my editorial, 'As we saw in the Child Welfare Exhibit.' I can count on a sympathetic, interested public." In Louisville, Kentucky, Strong was "amazed" by how newsboys took ownership of the child welfare exhibit in that city. They considered it "their exhibit" and contributed their own "enlightening information" to the exhibition script. Exhibits nurtured the "popular bias of thousands of people, once vaguely indifferent, now vaguely in favor, that urged the Providence legislators into their nine bills." Strong was not celebrating a "mob-consciousness," something she condemned, but observing powerful forces that she held in "far greater reverence, . . . the subconscious stirrings of the soul of the commonwealth, which is even now struggling into that conscious life which, when it is achieved, will be democracy."[9] This was the reformers' ideal—an enlightened democracy, in which the public is capable of learning and acting efficiently upon what they have discovered.

Child welfare exhibits gave children an opportunity to see themselves portrayed and to become conscious of their place within a larger and broader context. Strong took note of a fourteen-year-old boy at the Providence exhibit who stood watching the "vicious circle" that displayed the unbroken sequence of child labor, unskilled labor, low wages, poverty, and child labor. He said, "That means, doesn't it, that if I don't learn things, I'm going to be poor, and my children are going to be poor." For some children, exhibitions might involve a personal revelation. Strong observed another boy standing in front of an exhibit of "a bad home, dirty, ill-kept." Two other boys who passed by him expressed gratitude that "our house ain't nothing like that." A short time later, the first boy returned with a woman, and "for a time they talked in low tones," through which Strong heard him say, "But, mother, those boys said their house isn't like this. Why is ours?"[10] Raising consciousness among those on display was particularly gratifying for reformers. It affirmed their concern and their efforts.

Working in a seemingly universal language, reformers wholeheartedly embraced visual culture in the exhibit format as the ideal mode of communicating. Strong observed, "It is amazing to notice the extent to which charts and photographs are sprinkling public statistical reports." Strong maintained that "the immediate conscious purpose of the child welfare exhibit is, of course, not to legislate, nor to combine, nor to convert, but *to exhibit,* and by exhibiting to educate." The exhibit was "the answer to a great popular demand for easier and quicker ways of learning." As such, she argued, an exhibit was to be judged not merely by the legislation it produced but by a more sophisticated calculus that recorded "subtle changes of attitude and conviction, of individual and community relations, [through which] the child welfare exhibit works out its true purpose of popular education." The organizers of the Providence exhibit expressed their appreciation both to the National Child Labor Committee "for the services of Lewis W. Hine" and to his counterpart at the Playground and Recreation Association of America, Francis R. North, and for their work educating the citizenry.[11] In Rhode Island, Hine produced some of his most powerful work.

### PICTURING CHILD WELFARE IN RHODE ISLAND

The Rhode Island Child Welfare Conference and Exhibit was intended to "show in a graphic way the facts concerning child life." Reformers in the Ocean State had created a notable exhibit on the state industrial school for the Buffalo Exposition in 1901 and hoped in January 1913 to reproduce that success with a child welfare conference that brought together the entire community of child care givers. Through visual representations, reformers sought to show the entire community the range of problems faced by parents, teachers, ministers, social workers, and private citizens and to point toward their solution in existing services and the potential for new services. The scope of topics was broad, ranging from methods of dealing with dependent, "defective," and delinquent children to the juvenile court and probation system; the wider use of schools; the need for greater trade and vocational training; schools for the handicapped; infant care and mortality; dental and oral hygiene; housing problems; supervising amusements and commercial recreation; and other subjects relating to children's lives. Work was organized into five groups: education, recreation, health, the working child, and the neglected child.[12] Lewis Hine was deployed to provide a picture of existing conditions, the promises and the perils.

One of the outcomes sought by the Child Welfare Conference and Exhibit planners was the passage of laws proscribing the selling of newspapers by children under twelve years of age and prohibiting boys, and presumably girls, under the age of sixteen from selling papers after 8:00 p.m. The series of photographs that Hine produced for the exhibit in November 1912 were quite valuable in that they showed young children selling newspapers at all times of day and night and engaging in gambling, as well as newsies too young to be wandering the streets at all. Additionally, Hine recorded the names of the children in his captions, thus presenting them as specific examples rather than as types. Naming his subjects not only further humanized them; it also provided both reformers and historians the opportunity to look more closely inside particular children's households.

In the case of Irene Cohen (fig. 7.1), a ten-year-old newsie from Providence, Hine's photograph portrays a stoic young girl wearing glasses and standing with her hands by her side. The building foundation behind her creates a sense of size and scale. Hine actually took two photographs of Irene, once at medium range and once in close-up. In both instances Irene appears cooperative while looking expressionlessly into the camera. Irene was a Russian Jew who lived with her parents, Abraham and Florence Cohen, and her two younger siblings, Daniel (five years old) and Leonora B. (three). So many reformers objected to children working in the streets because they saw the urban environs as harboring the unknown, but Irene's father was a fruit peddler; perhaps his presence is outside the frame as protector or educator in the ways of the street.[13]

Rhode Island had had the highest levels of child labor among the New England states since 1790, when Samuel Slater launched the American Industrial Revolution in Pawtucket with a workforce of children. Hine photographed children at work in a variety of Rhode Island contexts. As long as the textile industry dominated the economy in Rhode Island, there were high levels of child labor. Hine did not do extensive survey work within textile factories on this trip, though he did take some photographs inside the textile mills in Lonsdale, Rhode Island. He photographed a mule spinning room and a warping room in the Lonsdale mills. Each image magnificently describes the occupation it depicts, whether mule spinner or warper. Each mule spinner tended between three hundred and one thousand spindles. In the warp room, we can almost hear the quiet and meticulous focus necessary to draw in the thread. Both photographs offer a clear representation of work in the textile factory.[14]

In his photographs for the Rhode Island assignment, Hine not only sought to illustrate exploitation or lost prospects; he also looked for opportunities to show pathways toward improvement. One of the areas in which reformers hoped to make legislative gains was the regulation of women and children's daily work hours. They argued that

FIGURE 7.1. "Irene Cohen, ten-year-old Newsgirl." (Hine 3162)

the U.S. government worked "able-bodied men only eight hours a day," so why did Rhode Island not "deal as humanely with its girls and women"? They also campaigned to lessen the number of hours worked on Saturdays, a reform that was debated in many industrial communities and often gained the support of merchants—"if it were binding on all their competitors as well." Girls were "standing fourteen hours," noted the program for the Rhode Island exhibit, and the resulting strain was "varicose veins, flat feet, injury to female functions." If women continued to be overworked, they were susceptible to "a nervous strain and over-fatigue which poisons permanently the whole system." Hine took two identical photographs in which the camera is focused on a seat attached to the end of a spinning frame or slubbing machine. The photographs were taken in a cotton mill in Pawtucket. In one, a female worker occupies the seat (fig. 7.2), and in the other the seat is empty (fig. 7.3). Like safety guards and other employee-friendly modifications, the addition of a seat at the end of the machine was an industrial improvement born of the need to address women's complaints. However, while the seat was appreciated by the workers, it was not so welcome when viewed as an alternative to a shorter workday.[15]

Hine photographed the first signs of the emergent corporate welfare approach to

FIGURE 7.2. "Cotton-mill employee using seat." (Hine 3192-A)

FIGURE 7.3. "Cotton-mill employee using seat" [*sic*]. (Hine 3192)

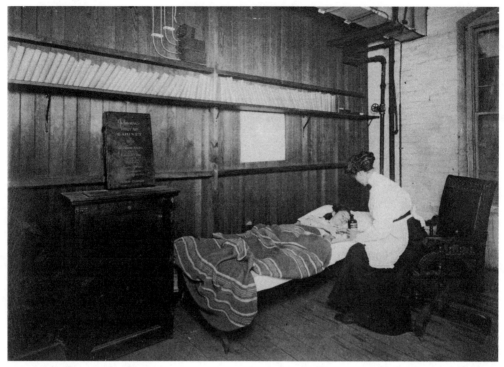

FIGURE 7.4. "A patient." (Hine 3186)

industrial relations. Pawtucket, Providence, Woonsocket, and Central Falls were no strangers to labor conflict, but rather than resist and regret every penny spent or lost, companies began to realize that investing in their employees through human relations and corporate welfare could be far more profitable in the long run. Companies such as Bryan-Marsh, an electric bulb manufacturer in Pawtucket, and Amoskeag in Manchester (see chapter 3), sought to improve employee relations and strengthen workers' loyalty to their employers by sponsoring company picnics, baseball teams, and housing, education, health, and welfare programs. The Rhode Island exhibit included Hine's photograph of a "sick worker receiving First Aid, in room set apart for hospital use and library" (fig. 7.4). Perhaps Hine's intent was to illustrate the dangers of industrial work, but the photograph may also be seen as displaying a new approach to changing industrial relations. The exhibit advocated that all children receive a medical examination from a physician to determine whether they were physically fit prior to receiving their working papers. The exhibit also brought attention to occupational hazards, urging the passage of laws in Rhode Island prohibiting children under eighteen from working in any occupation "declared by the state board of health to be injurious to life, limb or moral welfare," as twenty-four other states already had done.[16]

Overarching concern for the welfare of young workers brought benevolent surveillance beyond the workplace and into the intimate corners of child workers' lives.

Hine investigated living conditions in Providence and Pawtucket for the Child Welfare Exhibit and photographed working-class interiors and exteriors to display the powerful influence of the home on children's welfare. His photographs reveal the interiors of working-class living spaces to be modest, threadbare, and unkempt—perhaps unfit even to be called home. Sometimes the photographs showed calendars, crucifixes, and other attempts at individualization on the walls; often the images were stark and unsettling, such as in Hine's photographs of housing conditions on Elm Street in Pawtucket.[17] Did this pass as a "home" in reformers' parlance?

Hine's photographs of housing conditions were displayed in the home section of the Rhode Island exhibit. The conditions of the kind documented in his photograph of an "overcrowded home of workers" in the Olneyville neighborhood of Providence (fig. 7.5) were "strikingly shown through the good and bad flat exhibit," an immersion-type exhibit in which exhibitors created two home settings, each made up of a bedroom and a kitchen. One setting was shown in a state of "dilapidation and squalor," in keeping with Hine's photographs of Rhode Island workers' homes; the other portrayed the "clean well-kept conditions in which they might be maintained for the same money." Hine's photographs clearly showed conditions not as they might be but as they actually were—dilapidated and squalid. A problem with exhibits that juxtaposed good with bad was that, as one practitioner noted, "in many contrasting exhibits the 'bad' is often too bad to be convincing, and the 'good' too good to be true."[18] Hine's photographs possessed an aura of authenticity that helped to make that distinction less problematic.

Reformers believed that housing conditions were intrinsically connected to many health problems, and Hine's photographs provided examples of negative conditions from which health problems were sure to arise. Tuberculosis, or the "white plague," was a deadly serious bacterial infection, and exhibitors emphasized that housing conditions were a major factor contributing to the high tuberculosis rate in cities. In the broadest terms, reformers condemned "crowded and filthy dwellings [that] prove centers of infection of all kinds of diseases."[19] Hine produced photographs that showed these crowded conditions. His photograph of the Olneyville workers' home was supplemented by a caption telling us that eight people lived in three small rooms; in a rare bit of captioning, Hine also provided the size of the rooms. As Polish immigrants, the workers were believed to be endangered by the absence of Americanizing forces. The presence of boarders in those three small rooms intensified the health threat while also introducing the threat of immodest or immoral behavior. Hine's caption includes the address of the property, transforming it from an abstract type of dwelling into a specific house. On top of everything, Hine provided perhaps the most damning information: this was company-owned housing; the company that paid the workers' wages

FIGURE 7.5. "Overcrowded home of workers." (Hine 3191)

provided substandard housing to those same individuals. Hine gave reformers the pallet, or base map, from which they could broadcast their particular concerns.

Hine gave significant attention to the Italians living between Atwells Avenue and Spruce Street, an area known today as Federal Hill. Immigrants from southern Italy began settling in this central neighborhood in the late 1880s, and by 1910 it was overwhelmingly Italian. Hine's photographs of the area portray a neighborhood cut off from Americanizing influences, a neighborhood rife with examples of "otherness," of people clearly living outside the reformers' ideals. The Italians on Federal Hill understood home and experienced family life in ways that were different from middle- and upper-class norms. At a time when reformers were interested in confining family life within the home, the Federal Hill Italians were living externally, on stoops and in the streets. The boundaries between home and work that reformers sought to strengthen were largely absent within this immigrant community. Italian immigrants had not yet internalized or absorbed American virtues and values, at least in terms of living space.

The commonness of piecework being done at home on Federal Hill reflected Italian women's desire to stay out of the workplace as a matter of choice. As a result of this cultural preference, children became handy assistants in such homework. Judith E. Smith,

FIGURE 7.6. "Girls, six, nine, and eleven years old." (Hine 3172)

in her study of immigrant families in Providence, shows the complex calculations that went into taking in homework.[20] In each of Hine's photographs of homework, Hine situates his subjects in the center of the image, and the viewer's eyes immediately focus on them. Hine also captures enough of the interior space in each photograph to clearly identify the home as the context for the children's labor, and therefore for its condemnation. The "private" labors taking place within the house are exposed to the outside world. Absent are the sanctity of the family home and the segregation of work from home.

One of Hine's photographs of homework shows three girls, aged six, nine, and eleven years, as they work on chain-bags in their home on Knight Street in Providence (fig. 7.6). Providence was an important center of the jewelry industry, so constant work was likely available. The girls' parents, Mr. and Mrs. Antonio Caruso, came to the United States in 1898. The 1910 census lists Antonio as a weaver in a woolen mill. The girls in Hine's photograph are Anna, Emelia, and Gioviana. Not shown are the Carusos' other children, a four-year-old boy and a two-year-old girl. If they weren't already joining their sisters at the table around the time the photograph was taken, they soon would be.[21]

As we will see in a subsequent chapter, homework involved not just the labor

FIGURE 7.7. "Tiny girl with big bag." (Hine 3202)

performed in the home but also the transportation to and from the factories putting out the work. Among the most powerful of Hine's Federal Hill photographs is an image of a "tiny girl with big bag she is carrying home" (fig. 7.7). The juxtaposition of the large bag and the small girl creates a poignant reflection on homework and highlights the impropriety of little people doing big people's jobs. The girl's adult responsibilities are magnified when we realize that she is also providing child care and vocational training to her younger brother, who is running to catch up with her. The boy is symbolically falling behind. He is chasing childhood.

The immigrant working class that Hine captures in his photographs held family values that were different from those of the reformers. One of Hine's more interesting photographs of homework shows the "cigar factory of F. Delloiacono" on Atwells Avenue in Providence (fig. 7.8). Calling it a factory would seem to be quite the misnomer, as the image shows the Delloiacono family crowded into a small room that Hine's caption describes as "the living-sleeping-and-working room." Hine also notes that the room "adjoins the store" and that it is "very dirty and ill-kept." The setup displayed in the photograph demonstrates none of the segregation of uses that housing reformers sought. Relatively new arrivals to the United States, the Delloiacono family had

migrated from Italy in 1906, and they were still classified as "alien residents" in the 1910
U.S. census. All but the youngest children were born in Italy. The census shows that the
Delloiacono operation shared the building with five other families and their boarders.
The overcrowding and promiscuous use of space displayed in the photograph cried
out for reform of both working and housing conditions in the flat. One of the ironies
of the image is that cigar making was the only industry in which labor unions had
effectively ended child labor. The socialist leaders of the Cigar Makers' International
Union had waged a vigorous public campaign to encourage cigar smokers to buy only
those cigars whose ring bands displayed the union label, which testified that the cigars
were made without child or sweated labor. In this way they had put an end to the use
of child labor in large operations such as the R. G. Sullivan factory in Manchester, New
Hampshire. They had no control over some of the smaller shops in the tenements,
such as the Delloiaconos', but those were a disappearing breed.

Hine opened the doors to the unknown world of foreigners on Federal Hill. On Spruce
Street he photographed a bread delivery wagon (fig. 7.9). This wonderful image of two
men with loaves in hand, another man holding a bike, and a young boy slightly behind
them reminds the viewer that Italians, like other immigrants, sought out familiar people,

FIGURE 7.8. "Cigar factory of F. Delloiacono." (Hine 3177)

FIGURE 7.9. "Delivering bread in the Italian District." (Hine 3199)

culture, and language. This striving for the familiar provided ethnic entrepreneurs with an opportunity to fill the strong demand for traditional foods. Small wagon- or cart-based businesses sometimes grew rapidly as they nurtured a market for their goods.

Not only did immigrants bring their food ways from the old country, but they also brought the peasant mentality to the city. Hine photographed a young girl tending a goat in one of the back yards on Spruce Street (fig. 7.10). This image shows a child working, but not for wages or for any other remuneration. She is tending a goat that will contribute milk to the family. Cities often had to pass ordinances to prohibit such behavior. Pigs, chickens, and goats complemented the home gardens that were so popular among the immigrant working class, not only providing food on a nonmarket basis but also reinforcing more traditional patterns of family cooperation for family survival.

To fill out the unsettling portrait of otherness, Hine photographed three boys hanging around Spruce Street and smoking cigarettes, being delinquents (fig. 7.11). Hine made sure to note in his caption that they were "Italian Boys," thus contributing to the prejudicial notion that Italians were all involved in crime or were connected to the Black Hand, a criminal syndicate. Here again were the "embryo criminals," to use Scott Nearing's phrase (see chapter 2). Hine photographed a number of similar scenes:

FIGURE 7.10. "Back-yard, Spruce Street." (Hine 3198)

groups of boys lounging about, smoking cigarettes; truants; boys heading in the wrong direction, desperate for Americanizing influences to set them straight. Viewers were left to wonder which was worse, child labor or idleness. School or work seemed to be preferable to the total lack of oversight that gave the boys free rein in the city streets.

Hine's photograph of housing conditions on Republican Street on Federal Hill (fig. 7.12) magnified the sense of foreignness among the native-born. The image is of a man standing in a back lot with a group of children surrounding him. The wash hangs on a line overhead, bisecting the image. Paper and trash are scattered all about. There is no question whether this flat was bad. The children, like their surroundings, are dirty and unkempt. The image has two levels, so we observe not only the family in the trash-strewn yard but also an individual standing above them who appears to be directly taking part in the scene. Nothing or no one in this photograph seems to possess the new American values that reformers were promoting. Rather than showing parents and children indoors, this image shows them expanding their activities out of doors and into the lot behind their house. Instead of privacy, we have the intermingling of families in a promiscuous space, where children in particular are vulnerable to strangers cutting through the lot. Hine the photographer stands above both of the image's

FIGURE 7.11. "Cigarettes, Spruce St., Italian Boys." (Hine 3201)

levels, his shadow cast down, making judgment upon these conditions. American housing reformers were at odds to find anything positive in this image, the details of which violated so many basic ideals.

Rather than being viewed as a description of family degradation, this image can be read as an affirmation of family—not a neat and clean family, but an expansive one that exists beyond its particular flat, overflowing into the street. The adult male in the photograph, presumably the father of all the children, holds a young child in his arms, looking comfortable and in control. There are no adult women in the image, though one can easily imagine that this lot was where laundry was done and where people, their "own kind," gathered. Still, while the photograph might reflect the realities of Italian immigrant life, it does not match any of the goals for improvement in child welfare.

Hine created numerous portraits of dirty neighborhoods in which health problems awaited their victims. He photographed the rear of a tenement on Spruce Street that was the entrance to a midwife's house (fig. 7.13). Hine's caption describes the house as crowded and dirty, though he needn't have mentioned the filth. The image is powerful for several reasons. Consider that women came to this highly unsterile environment to deliver their babies. This photograph begins to explain why infant mortality rates

FIGURE 7.12. "Housing conditions, Rear of Republican St." (Hine 3203)

in textile cities such as Fall River, Lowell, and Manchester were among the highest in the country. People other than expectant mothers also sought medical assistance in the midwife's home. Suspicious of hospitals as places to die, they turned to alternative institutions that failed to meet the health standards that were being developed and were desired by reformers. This was one reason for the establishment of the Visiting Nurses Association, which could bring health education and assistance to the neighborhoods. The small boy standing alone near the ash pile seems to be dwarfed by the filth around him. The health threat is obvious.

Hine also photographed the front of a midwife's home on Spruce Street (fig. 7.14). Fifteen small children gather at the entrance, one adult woman stands behind them, and a slice of another face appears in the doorway. Was this the neighborhood health center? Was this the only institution in the neighborhood actively promoting child welfare? Hine was a master of the small detail that authenticates an image. On the corner of the house in this photograph, just above and to the left of the door, hangs a sign that reads "Cipolla Levatrice Italiana Laureata nell'Università di Napoli,'" which translates as "Cipolla Levatrice, Italian, graduated from the University of Naples." Counter to the stereotype of uneducated immigrants, the midwife advertises herself

FIGURE 7.13. "Entrance to the crowded, dirty house of a Midwife." (Hine 3170)

FIGURE 7.14. "One of the Midwives' homes." (Hine 3200)

FIGURE 7.15. "Housing conditions, Woonsocket, R.I." (Hine 3160)

as a university graduate. The world of the new immigrants was unknown and foreign to many who sought to transform and uplift that world.

Hine photographed housing badly in need of repair or of demolition and removal, even if not an immediate health threat. The run-down conditions that Hine exposed not only blighted the neighborhoods; they also were host to unsupervised places that were dangerous for children to play in. In a photograph of the Stone House on Hill Street in Central Falls, Hine situates two small children carrying lunch pails at the center of the image; they are almost buried by the rubble that surrounds them. Does this image in some way portray a vision of the end of the world, suggesting that civilization will crumble if children labor? Probably not, but it does show decomposition and the blighted neighborhood that young children traversed on their way to work.[22] Hine created a similar image in Woonsocket, portraying decaying housing conditions that were breeding grounds for a diseased and disorganized family life (fig. 7.15). At the edge of this cliff of decay is the silhouette of a small girl, a symbol of life within a dying group of buildings.

Hine escalated reformers' anxieties about unsanitary urban conditions with a series of photographs that showed neighborhood privies—nasty, indiscreet, and almost certain health risks. His photograph of the back yard and privies of a house on Borden

FIGURE 7.16. "Back-yard and privies." (Hine 3218)

Street (fig. 7.16) brought together multiple themes of urban life, amplifying the need for reform. Hine's caption tells the viewer that the privies are in "terribly filthy condition," though again the image "speaks" to the same point.

Having presented the problems, Hine also photographed those engaged in the promotion of child welfare. Sprague House was started on Atwells Avenue on Federal Hill in 1910. Hine's photograph of the exterior of the Sprague House Settlement on a Saturday morning shows seven young children climbing on the porch railing as an adult couple strolls by, the woman holding a child in her arms (fig. 7.17). A man and a woman with child in tow—that was more like what families should look like. Hine gives no hint of what is going on inside the building, but he does caption that the "house is so full that these children could not go in." Was Sprague House popular? Why? What types of services and programs did it offer? The image provided a starting point for the new settlement house to define itself more fully within the constellation of child welfare reform organizations by handing out or displaying supplemental materials at the conference or exhibit.

Hine made a second photograph of the Sprague House Settlement that is remarkable for what it shows going on inside (fig. 7.18). In this image a sewing class of nine young girls sits in a semicircle open to the camera; each child is stitching a small piece

FIGURE 7.17. "Exterior of Sprague House Settlement." (Hine 3175)

FIGURE 7.18. "Sewing class in Sprague House Settlement." (Hine 3174)

of cloth. Against the wall is a piano, its keyboard closed for now, but honored by its central place in the room. The director of the settlement also appears in the photograph; she is standing behind the girls, holding what Hine's caption identifies as "a newly-arrived deserted baby." A portrait of a woman holding and comforting an infant hangs on the wall opposite the director and acts like a mirror to her. The children don't seem startled by the new arrival—they are all intent on their sewing—but for the viewer the revelation invites new reverence for the scene. If there was any doubt about the importance of the settlement movement, this photograph was its refutation. The Sprague House Settlement was a baby-saving institution.

## BETWEEN THE PROFESSIONAL AND THE PUBLIC

The popularity of exhibits addressing social welfare issues was never higher than in the first two decades of the twentieth century. Child welfare exhibits; baby-saving exhibits; tuberculosis and pure milk exhibits; industrial exhibits; recreational, play, housing, or planning exhibits—all fit the need of the new professional class to communicate specialized information to a previously indifferent public. In their book *The A B C of Exhibit Planning* (1918), Evart Grant Routzahn, associate director of surveys and exhibits for the Russell Sage Foundation, and Mary Swain Routzahn articulated a widely held belief that "information on social welfare is growing rapidly, and we must close the gap between the small group of socially informed people who keep abreast of this knowledge and the great mass of those whose understanding and co-operation must be gained before the application of the knowledge can be made." Exhibits were a means of transmitting professional knowledge to the public mind. The Routzahns maintained that the exhibit had "already played an important part in closing this gap, particularly on the subjects of public health, child welfare, and the care of certain groups of people who have become dependent."[23]

Selecting the appropriate photographs to exhibit was exceptionally challenging, and not everyone who tried did it well. Exhibitors using photographs increasingly sought to enlarge the images to achieve the greatest effect. The design of the panel was important as well, to make the most of the information being communicated. Evart and Mary Routzahn called for a heavy hand in editing each image, where "all irrelevant matter should be cut away so that the significant features of the picture stand out boldly. Either the text of the panel or the label for the photograph should bring out unmistakably what idea or facts each illustrates." They edited not necessarily for truth but for influence. The Routzahns continued, "Many examples of the possibilities of manipulating the details of photographs to obtain striking results or to make a picture fit into a particular space or design are afforded by familiar newspaper practice." They stressed that it was not enough to select an image because it was attractive or remarkable. Photographs needed to elucidate meaning and message rather than be mere decorations or visual litter cluttered with conflicting evidence; they had "to make the exhibit more easily understood."[24]

By 1915, Hine was recognized as one of the masters of exhibiting photographs successfully. The Routzahns cited a photographic panel by Lewis Hine for an exhibit of the National Child Labor Committee as an illustration of the way in which photographs can "make the message of the panel vivid." Hine was exceptional in his ability to produce photographs honed to the message. Not everyone was so talented, of course, and poor selection of photographs could muddle the meaning and destroy the effectiveness of a panel or an exhibit. Additionally, the use of photographs in social science exhibits began to decline by the end of the 1910s. According to the Routzahns, to an "increasing extent cartoons and free-hand sketches are being substituted for photographs because of their ease in bringing out the real points of the illustration and freedom from needless and distracting detail."[25]

It wasn't always possible to bridge gaps in the population between reformers and reformed, or to unite the divided interests of an urban community. Exhibits did not always bring people together; they could be competitive as well as cooperative and generate conflict over the visual record. In one instance in North Carolina, textile manufacturers created their own exhibit to challenge the impressions of an NCLC exhibit. The *Charlotte Daily Observer* reported in January 1911 that the National Child Labor Committee had set up an exhibit to run for several days in Raleigh while the North Carolina General Assembly was considering important child labor legislation— the prohibition of children under fourteen years of age from working in the cotton mills. The state's cotton manufacturers were "not asleep," however: not to be outdone, the "mill men" were arranging an exhibit of "the actual conditions as they really exist to counteract any possible erroneous impression" made by the NCLC exhibit, which they said would focus on "isolated and exceptional cases as they are usually shown by the opponents of child labor, which may be found in every industry in the world no matter how well regulated." When only the exceptions were held up for public scrutiny, complained the mill men, "that is where the great hurt to the industry is done."

The cotton manufacturers planned to include images of working conditions in Greensboro, Raleigh, Charlotte, Kannapolis, and elsewhere that would "represent the general status, the average of the industry." They believed that showing "illustrations from cities and towns familiar to the members of the Legislature" would go a long way toward countering the "erroneous impression" given by the NCLC exhibit. The mill men claimed that they were not opposed to raising the age limit for child laborers, but they wanted to do it gradually so that "the parties most concerned are ready for it." They advised that school attendance must be ensured for the children once they were no longer working; otherwise the children would simply "idle away their time" and would actually be harmed more than if they were "doing light work in the factories." The NCLC exhibit, they charged, would "doubtless be a pitiful wall, with figures and statistics, et cetera, ad libitum, and photographs depicting something of the awful conditions that are said to prevail."[26] The cotton manufacturers did not want the bad publicity, or *any* publicity they could not control.

In the summer of 1915, the *Boston Globe* reported that the Boston Industrial Development Board had declined to participate in a popular series of free evening entertainments provided for adults in the parks and playgrounds of the city. The programs for these "park shows" were a mix of news pictorials, comedies, cartoons, and health and social welfare features, pictures, and slides; the board objected to the inclusion of "certain lantern slides dealing with the child labor question," which they felt should be omitted.[27] The board believed that Massachusetts should "halt the agitation for changed labor laws" until other states had passed legislation more in line with Massachusetts's existing child labor laws. In a letter to the Committee on Park Shows, board chairman John N. Cole wrote, "I have seen the Child Labor laws made more stringent every year since I began work as a boy of 13 years of age, until today Massachusetts actually leads the country" in protecting child laborers. Consequently, photographs and "statements as are made relative to this subject are not justified by existing Massachusetts conditions." Given the many people who were "naturally dissatisfied with conditions," Cole continued, the images would "serve only to arouse class feeling, resulting ultimately in real injury to the particular class of people who are supposed to be most benefited by such exhibitions as these Summer shows provide." Cole argued that there were enough subjects "to make interesting programs without dragging in the harassing negative side of life, such as the pictures referred to undoubtedly do." As such, he concluded, the Boston Industrial Development Board could not in good conscience cooperate with the park shows and contribute money "for a propaganda which includes an agitation for aroused class feeling."[28]

The Committee on Park Shows, which represented a group of ten private organizations interested in health, recreation, and community welfare, subsequently met to consider how best to respond to business opposition to exhibiting child labor. To be effective in the world of Progressive reform, organizations had to appear to be nonpartisan, to be using their expertise as their rationale for action rather than following a partisan agenda. To counter charges that its program selection was prejudicial, the committee wrote to Cole, the chairman of the Boston Industrial Development Board. Asserting a completely "impartial attitude," the committee reminded Cole that it had recently shown "motion pictures" supplied by the National Association of Manufacturers that examined industrial safety devices and "other desirable conditions in manufacturing."[29] Nevertheless, the exhibition of materials that exposed social conditions threatened Cole's board, and even appeals to balanced displays failed to move the business group.

The committee then addressed the fundamental question at hand: Were these photographs truthful, and if so, did they warrant display? The committee informed Cole that it had received "no information to justify discontinuing the use of its child labor slides." Indeed, as Lieutenant Governor Grafton D. Cushing, chairman of the Massachusetts Child Labor Committee, stated in a letter to the Committee on Park Shows, "the slides show actual photographs of Massachusetts conditions which still exist." Therefore, he argued, it became the committee's duty to see that "the public should be informed of conditions which are still open to improvement." The problem in Massachusetts, Cushing noted, was less with the laws, which were "fairly satisfactory," than with the fact that "they are not yet completely enforced."[30] The park show went forward as planned, but the battle for control of visual culture was just beginning.

## CHAPTER 8

# Homework

———

IN A 1914 BULLETIN ISSUED by the National Child Labor Committee, Lewis Hine described homework as "one of the most iniquitous phases of child-slavery that we have."[1] The Massachusetts Bureau of Statistics defined industrial homework as "the manufacture or preparation within the home of goods intended for sale, in which the work supplements the factory process."[2] Massachusetts officials had tried to regulate homework since 1891, but the practice continued and flourished into the first decades of the twentieth century. Homework transgressed reformers' ideals of family and of home, two of the most idealized social arenas of the period. It erased the boundaries between residence and workplace and brought the manufacturing economy into the domestic space. In repeated investigations of homework, the same "attendant evils" were identified—long hours, low wages, unsanitary conditions, and the use of child labor.[3] Reformers felt strongly that the home should be preserved as a private place for the enjoyment of family life.

The historian Susan Porter Benson has argued that homework in New England not only was at variance with the sanctification of the home but also was a social problem. Homework first and foremost was exploitative of women and children, the most vulnerable groups in society. Homework depressed wages paid in factories and undermined union efforts to improve wages and conditions. And because homework was conducted beyond the reach of factory sanitation rules and regulations, minimal as they were, it added to the unhealthfulness of the private and public lives of the working class.[4] On the other hand, while homework was widely condemned for mothers, it was a way to keep children close to home. Otherwise they were likely to be set loose in the city streets, since kindergartens and nursery schools were in their infancy and unavailable in most communities. Homework also allowed children to contribute to the family economy, which was particularly important among ethnic groups in which the cultural prohibitions against working outside the home were strongest.[5]

The "evils of the work" were plain for those who wished to see, according to the Massachusetts Child Labor Committee. Industrial homework was done in "poorly-ventilated rooms, often in dirty kitchens and in unhygienic houses." Such conditions harmed the health of the child worker and were "dangerous to the health of the community." When a child with ringworm was "found crocheting underwear," or work was done in a household in which "a contagious disease was present," or the piecework was "dragged about in the dirt and crawled over by the babies," the threat of homework was not just to the child or the family: it was perceived to be a social threat as well. The long hours and late nights spent bent over their work, struggling to see, caused children to develop "nervous strain" and interfered in their schooling. The "anemic, tired, nervous, overworked children" were driven "until they [cried] out against the abuse." Reformers believed that the children were uncared for and that their playtime was being "stolen" by their labors. A child so raised, reformers feared, would soon "realize the exploitation he has suffered[;] he will mourn those hours wasted in uneducative monotony, he will mourn that heritage of physical and mental strength which has been bargained away for a pittance."[6] Reformers, Lewis Hine among them, set out to gather evidence of these evils, of family pathologies that left unattended would soon threaten both the health of society and the social order.

## LEWIS HINE ENTERS THE HOME

Hine's investigations into homework required that he enter the intimate spaces of the working class, and the resulting photographs exposed those spaces as falling far short of middle-class ideals of home and family. The home as a site of industrial production quite naturally became a target for inspection, with working-class and immigrant households ultimately subjected to new levels of intrusive government surveillance as well as the attendant observations and judgments of reformers. Bringing reformers into working-class homes and allowing them to critique the lives and living conditions of immigrants and the working class was a significant expansion of the reformers' gaze into the private world of family. Entering a home required Hine to gain the cooperation of the family, since he presumably did not have the power to enter homes where he was not wanted. Families, therefore, had to consent to the inspection and to his taking photographs. Hine had to establish a tentative relationship with his subjects to get them to focus on their work while he focused his camera on them. His captions brim with details that evidence the discussions that took place.

Of all the child labor settings Hine photographed and made surveys of, he held

particular disdain for homework. In New York and elsewhere, Hine did considerable investigation of homework. The subject, he told an audience at the 1914 National
Child Labor Conference, brought him "nearest to hysterics," evoking "memories of
twenty-four hour shifts tripping up and down the tenement stairs, six or seven flights
of them (for the good things are always at the top), loaded down with several tons of
camera equipment," and of numerous times in which he was very near to "getting what
has been long coming to me from those who do not agree with me on child labor matters." In New England, Hine photographed many instances of homework that threatened family life, all of them contributing to his condemnation of the practice. He told
his audience, "If I seem to be smiling over the subject at any time this morning, you
may rest assured I would rather weep."[7] Hine expressed greater personal indignation
over homework than he did over any other type of child labor.

Hine was incredulous that manufacturing had turned the home into a workshop
and transformed family members into a divided and subject labor force requiring parents to take on unwarranted overhead costs and to assume the disciplinary role of
management.[8] The market worked its way into the homes of the working class more
through homework than through any other form of industrial labor. "Never before
on so great a scale have working people's homes been invaded by industry," observed
Florence Kelley, leader of the National Consumers League and a principal advocate
against child labor. Homework tore away the veil of family sentiment and revealed the
home to be not a haven from the harsh industrial system, as middle-class reformers
would have it, but a particularly exploitative place, precisely because it violated the
boundaries that, for middle-class families, preserved home as a place of respite from
the harsh terms of capitalist production. Middle-class family ideals clashed with the
expansive character of capitalism and its incessant need for profits. Reformers inserted
themselves into the exchange to moderate the aggressiveness of industrial capitalism,
particularly as it applied to the home, women, and children. Of course, most of the
working-class families that Hine photographed could not yet afford those separate
realms, even if they wanted them.[9]

" 'Be it ever so humble, there's no place like home'—for what?" Hine sarcastically
observed in his address at the child labor conference. "Why be selfish and try to
reserve the home for the family?" Continuing to speak facetiously, Hine appealed to
his audience to follow the path blazed by American business. "Industry," he explained,
"with its usual unselfish, altruistic spirit, has been demonstrating for years how we
may utilize waste, and everyone knows how all the industrial by-products are turned
to account." He noted that "the master minds of industry have been greatly disturbed
over the forces and activities in the home that, like the mighty Niagara, are sweeping onward and accomplishing nothing." They saw "mothers spending whole days at
housework, children at play, neighbors visiting with each other. But," he asked, "what

does it all amount to?" Continuing his mock critique, Hine wondered aloud whether "these golden moments could be harnessed to the wheels of Industry" and asked his audience to imagine each of the nation's 25 million children knitting "for three hours a day (before and after school, of course)"; they would be able to "keep the world in socks"—or if their work time were doubled to six hours a day, "they could make our nation's shirts." Families would increase their chances of making a "starvation wage if every member of the family worked all the time," charged Hine. The virtue of this plan kept "Johnnie and Jennie fully occupied when they might otherwise be wasting their time in fruitless play." Cynicism was not enough to combat the encroachment of child labor on the home.

Mocking the old Puritan ethos, Hine challenged his listeners: "If the little hands are always busy how can Satan find mischief for them to do?" In a society where the business of the nation is business, he argued, it shouldn't be surprising "that our children must be reared in an atmosphere of work if they are to become Captains of Industry, and so we outdo the Montessori System itself, finding tasks for the tiny tots to try." Hine then pretended to correct himself, positing that the various types of homework children were engaged in—making garments, sewing on clothing labels, setting stones in jewelry, rolling cigars, crocheting, stringing lace, making flowers, and so on—were not "tasks" but instead were "'opportunities' for the child and the family to enlist in the service of industry and humanity. In unselfish devotion to their home work vocation, they relieve the overburdened manufacturer, help him pay his rent, supply his equipment, take care of his rush and slack seasons, and help him to keep down his wage-scale."[10] Others shared Hine's societal assessment. For instance, Mary Van Kleeck, secretary of the Committee on Women's Work of the Russell Sage Foundation, wrote that it was "characteristic of the system [of homework] that its evils are its life, that the unrestricted competition of unskilled workers, the unregulated hours of work, and the employment of children are the things which make it profitable for the manufacturer."[11] The manufacturer alone seemed to enjoy those profits, as workers' wages were notoriously low.

## THE UNPROTECTED CHILD AT HOME

Children toiling in the home were the least-protected laborers. By the early part of the twentieth century, the majority of states had passed some form of child labor legislation, but most of these laws concerned only child workers in manufacturing. Even the "weakest child-labor law," according to the muckraking journalist and suffrage advocate Rheta Childe Dorr, was at least "a declaration that manufacturers may not employ children under a given age without permission from the state, nor may they employ children outside the terms fixed by the state." Such legislation stated clearly that "no parent has a right to send his children into industry except under conditions defined by the state." Homeworkers were not so fortunate, however, for "a parent may claim the services of his children from the day their baby hands are capable of performing a task. He may work them until they drop asleep from exhaustion. He may put them at work injurious to health, in surroundings actually conducive to physical and mental destruction. Provided the work is done at home."[12]

The sanctity of the home within American culture inhibited effective action at the same time that it called for such action. Dorr observed that "this remarkable distinction which permits people to do to their own children that which the law prohibits them from doing to the children of others, is a survival of the old sentimental theory of the innate sacredness of The Home." Reformers such as Dorr, however, contended that "a child who does factory work at home needs the same legal protection as the child who does factory work in a loft building." Studies made by various committees and organizations concerned with social welfare had proved that the home factory was "neither a home nor a factory, but a miserable sweat shop. *Nowhere else,"* Dorr maintained, *"are child-labor laws so desperately needed."*[13] No more intimate workplace existed, and no type of work needed exposure and legal protections more than homework.

Reformers seeking to protect homeworkers of any age had essentially two paths open to them—the outright abolition of homework, which appeared an unrealistic goal, or government regulation, empowering state intervention into this sacred realm. Either choice, to abolish or to regulate, meant an expansion of governmental authority into the most intimate of places. Reformers believed that homework threatened society by contributing to ignorance, ill health, and economic exploitation, and that in every way it undermined the growth of the rising generation. Rather than trying to reform society, however, they pathologized the families they interviewed. The historian Linda Gordon has argued that, "although this pathology might be seen to be rooted in poverty, racial or ethnic inferiority, and the degradation of urban life, the Progressive-era child protectors were mainly drawn to the opposite causal conclusion: that family weakness was at the root of larger social problems." Efforts to ameliorate such problems, Gordon maintained, "were always aimed at reforming families, not society."[14] This was the approach Hine took as well in exposing homework, highlighting individual family issues rather than larger social or economic ones.

Organized labor, long an opponent of homework, attempted to use its growing strength to prohibit the practice. From the time of the union's formation in 1891, the United Garment Workers of America took a forceful stand against homework, calling for it to be outlawed "on the ground that it is detrimental to the health of the workers and lowers the standard of wages in the trades concerned." The Cigar Makers' International Union was perhaps most successful in combating homework, employing the union label to signify that a cigar was free of sweated labor. A Massachusetts survey discovered that "cigar makers have practically stamped out home work upon tobacco, except for independent manufacture in tenements, which is not regarded as home work." Following the lead of the Cigar Makers' Union, the National Consumers League began to affix a "Consumers League Label" to all goods made by—that is, on the premises of—a manufacturer.

Periodically, union laborers struck to prevent homework from further infecting and eroding their trades. In the summer of 1910, there was a strike among cloak, suit, and skirt makers in New York in which the workers won a protocol agreement stating that "no work shall be given to or taken to employees to be performed at their homes." In the spring of 1913, Boston garment workers also struck, in part over homework. The settlement of that strike included agreement on the abolition of homework and of subcontracting between employees. The success of those negotiated agreements is unclear, and the success of the label campaigns even more so. However, by looking for the cigar union label or the Consumers League label, union workers exercised their collective power as consumers to boycott exploitative practices, abusive manufacturers, and sweated goods.[15]

Although many reformers favored the abolition of homework, regulation seemed a more practical and more popular approach to the homework problem. By 1914, twelve states had enacted legislation regulating homework: Connecticut, Illinois, Indiana, Maryland, Massachusetts, Michigan, Missouri, New Jersey, New York, Ohio, Pennsylvania, and Wisconsin. Eight of those states required the licensing of places in which home manufacture took place.[16] In New York the law provided for the licensing of all homes and tenements in which any one of some forty articles were manufactured. Still, conditions were worst in New York, according to Rheta Childe Dorr, who reported that 9,805 tenement houses in New York City were licensed for homework in 1908, adding that it could not be known "how many *unlicensed* tenement houses contained factories." By 1911 there were more than 14,000 licensed tenements in the city. The problem, noted Dorr, was that "in these houses, and in nobody knows how many more, live from three to thirty families, and instances are multiplied in which every family in a building was engaged in home manufacturing."[17] Mary Van Kleeck maintained that the history of New York's attempts to protect homeworkers illustrated "an attempt to 'regulate' a system which thrives on failures to regulate it."[18]

## INVESTIGATING HOMEWORK IN NEW ENGLAND

Despite its weakness, regulation was the only valid path of reform, but successful regulation required accurate information, the acquisition of which quite often depended on survey work and home inspection. In New York City, investigations into homework were carried out by the state labor bureau, the national and city child labor committees, the national and local consumers' leagues, and the College Settlements Association.[19] Elsewhere, the story was much the same. In 1914, the Massachusetts Bureau of Statistics published the results of an investigation into homework that had been made in conjunction with the Women's Educational and Industrial Union's Department of Research. The goals of this investigation were to identify the industries in Massachusetts in which homework was found, the influence of homework on factory work and wages, the types of families engaged in homework, their motivation for working at home, and the nature of homeworking income and its effect on family life.[20] Lewis Hine, with extensive experience investigating homework in New York City, joined the Massachusetts investigation to gain evidence for both the national and state child labor committees. Other surveys were conducted in New England in addition to the Massachusetts survey. In 1919, the Women's Bureau and the Children's Bureau of the U.S. Department of Labor undertook a survey of homework in Bridgeport, Connecticut. Years later, the Women's Bureau conducted a survey of homework in the lace industry in Rhode Island.[21]

Women played a leading role in these surveys, both as investigators and as the subjects of investigation. Their predominance in the survey teams studying homework reflected women's expanding commitment to "municipal housekeeping" and their critical place within the emerging profession of social work. The Department of Research of the Women's Educational and Industrial Union initiated the idea for the Massachusetts survey and assembled a talented team of investigators, enlisting three of the Union's research fellows—Margaret Hutton Abels, Margaret S. Dismorr, and Caroline E. Wilson—to serve as field agents for a period of nine months. Dr. Amy Hewes, professor of economics at Mt. Holyoke College and secretary of the Massachusetts Minimum Wage Commission, was asked to oversee the investigation. Many other women contributed to the survey over the course of the eighty weeks spent in the field and during the subsequent period of analysis and work on schedules. (The only male participant named in the final report on the inquiry was Frank S. Drown, chief statistician of the Labor Division of the Massachusetts Bureau of Statistics; he helped supervise the fieldwork and prepared the tables for the report.)[22] The Bridgeport survey, initiated at the request of "local agencies interested in industrial conditions," was overseen by Mary Anderson, director of the U.S. Women's Bureau. Anderson also directed the Rhode Island survey, which was conducted by Harriet A. Byrne and Bertha Blair.[23] These women and others like them expanded the social responsibilities of the state through their social feminism, pushing local and state governments to intervene when children's well-being was in the balance. Professional women applied their expertise to the regulation of children's labor in the home and in doing so significantly expanded the reach and responsibilities of the state.[24]

In 1912 Hine, sometimes working with F. A. Smith of the National Child Labor Committee, surveyed homework conditions in Massachusetts, taking dozens of photographs of homeworkers, primarily in Boston, Somerville, and Cambridge but also in Northampton, Easthampton, Leeds, and Worcester. Although Hine's survey work and the three southern New England surveys varied significantly in scope, their findings were, with some notable exceptions, fairly consistent with each other. As Mary N. Winslow wrote for the Women's Bureau, "Investigations of home work and its attendant evils have been made many times, and the findings have been practically identical. The evils of home work—low wages, long hours, insanitary conditions, and child labor—are no recent discovery; but the problem is still a matter of discussion, as no solution has been found."[25]

The Boston metropolitan district was the only area of Massachusetts in which the licensing and inspection of homework was carried out, and only those families that worked on wearing apparel were subject to regulation. Other Massachusetts towns that produced wearing apparel—Chicopee, Northampton, Foxboro, Haverhill, Leominster, Newburyport, Reading, Salem, Framingham, West Springfield, Springfield, and Worcester—were free from similar oversight. Homeworkers around the state produced many different products, and reformers believed that it was important to regulate other home industries in addition to apparel. Families worked on a variety of popular paper goods: frills, skewers, boxes, plates, paper napkins, paper doll outfits, flowers, rosettes, fans, caps, circulars, envelopes, and favors. Celluloid goods such as fans, bandeaux, napkin rings, boxes, and cards and nests for hairpins were also common products of home industry, as were brushes (including toothbrushes). Silk goods, including darning and embroidery silks; curtains, bedspreads, and dresser covers; and centerpieces, doilies, towels, table and bed linens, and handkerchiefs were all partially produced in the home. Homeworkers also made jewelry and silverware, sporting goods (including fishing rods and all types of balls), and toys and games. They sometimes even worked with human hair.[26]

The issue of homework was by no means parochial. Home industries were the subject of study in numerous countries, including Britain, France, Germany, Italy, Belgium, the Netherlands, and Finland. The House of Lords, appointed to investigate the "sweating" system in Britain, reported in 1890 that the "evils" of homework—inadequate wages, long work hours, and unsanitary working conditions—could "hardly be exaggerated." In 1908, the House of Commons appointed a Select Committee on Home Work, whose subsequent report lamented "the almost complete absence of statistics on the subject" but offered no evidence itself beyond the testimony of various witnesses before the committee. The committee's report did proffer two remedies: the creation of wage boards to fix and adjust minimum time and piece rates, and regulation and inspection to supplement the actions of the boards. In France, the Bureau of

Labor conducted investigations of homeworkers in the lingerie industry, in the artificial flower industry, and in the boot and shoe industry.[27]

While it was nearly impossible for the New England surveys to determine definitively the extent of homework within their respective regions, they all shared certain conclusions. Chief among these common findings was that homework was a family affair and child labor was central to that work. Homework was widespread in the region and was particularly common in working-class and immigrant neighborhoods. As the Massachusetts report stated, homework could be found anywhere that "industrial establishments, with a product upon which outside work can be done, have gained a foothold." Wherever investigators looked, they found that child labor was a "conspicuous evil associated with home work."[28] Amy Hewes, commenting on the inconsistencies in the popular mind-set, noted that "young children, long ago forbidden by law to work in a factory, and women, usually prevented by domestic duties from engaging in regular industrial occupations, make up the greater part of the labor force."[29] Rhode Island investigators, who noted a similar prevalence of children among the homeworkers they surveyed, concluded pessimistically, "The prevention of child labor . . . is practically impossible under a system of industrial home work. It is a temptation in families that eke out their existence by home work to increase their pitifully small earnings with the aid of even very small children."[30] Investigators for the Bridgeport survey visited one hundred families and counted 268 children under sixteen years of age; 110 of those children "were definitely shown to assist regularly in home work and it is probable that others in the group are also helping." As one mother in Bridgeport plainly told investigators, "Home work isn't worth bothering with if the children don't help."[31] Children were vital contributors to all aspects of homework.

## HINE PHOTOGRAPHS THE HOMEWORKERS

Hine photographed numerous families engaged in one or another form of homework; children have a prominent place in these images. By and large these photographs show families working together, though usually without the participation of an adult male wage earner. Homework fell entirely within the female dominion, and women governed the operation of homework in the household, deciding who would work and when, as well as how and for what (if any) remuneration they would work. In photographing the home of Rufine Morini in South Framingham, Massachusetts (fig. 8.1), Hine built the image to highlight the idea of the family as the unit of production, with children as vital contributors to the family economy. Hine photographed the family sitting around a table piled high with the product of their labors, the value assigned by an unseen market and by the desire of their employer to get more and pay less. Hine captions that there are "two mothers" and three children at work. The family continues

Figure 8.1. "Home of Rufine Morini, 6, Coburn Street." (Hine 3140)

to labor in spite of the presence of the father, who stands holding an infant. He appears an interloper, ready to go out the door and leave them to their work. Or perhaps not. Viewers can make that determination for themselves.

Low wages were also characteristic of homework, whether or not the children helped out. Hine's photograph of Mrs. Donovan's Roxbury household documents a family that had been tying tags for the Dennison Company for seven years (fig. 8.2). Hine photographed the mother and children working outside the home; they are sitting on the porch or perhaps the back stairs of their building, with Mrs. Donovan perched on the highest step, supervising her young laborers, who were all between four and a half and thirteen years old. Even with all the children participating—and often having "to work late at night to get done"—the household's average monthly income from homework was only thirty dollars. One month they earned forty-two dollars, but Mrs. Donovan questioned whether they would "ever be able to do it again."[32] The area around the Donovans is starkly bare, save the odd rag, the broken fence, and the run-down walkway. The children's dirty bare feet reinforce the message that these young workers were earning low wages.

The investigators for the Massachusetts survey offered as a universal truth their

Figure 8.2. "Family of Mrs. Donovan, 293 1/2 Highland Street." (Hine 2982-B)

conclusion that "a low average of wages generally prevails for home work." They found that 59.5 percent of those who had received payments for at least nine months during the preceding year had earned less than $100 for the year; 78.5 percent had earned less than $150; and only 4.1 percent had earned more than $300. Half of those who reported their hourly earnings made less than eight cents an hour, and 22.5 percent earned less than five cents an hour.[33]

In his caption for each homework photograph, Hine sculpts a brief biography of the household as its members gather around their work. Each image is coupled with a story that tells the viewer who was doing what work, where they lived and labored, the number of children and their ages, and the family's earnings, along with some added comment on their particular circumstance: a tale of exploitation, a threat to their health, a glaringly low income, a loss of educational opportunity. Each image contains the elements necessary to evoke interest in reform—not necessarily among the home-workers themselves, but among their sociological "others" residing in homes purged of commercial life. Hine's captions closely resembled the information gathered by social workers to form family case files. Each story stood alone, a case whose singular experience, when combined with similar family cases, posed a threat to social order by displaying the common ingredients of ignorance, poverty, and disease that confronted each family and therefore society as a whole. The photographs created by Hine became part of the case studies. They formed an archive of alarm for those concerned for the coming generation of Americans.[34] They also reflected the role played by Hine in the process of modernity, in which his photographs are, as John Tagg argues, "bound up with the emergence of institutions, practices and professionalisms bearing directly on the social body in a new fashion, through novel techniques of surveillance, record, discipline, training and reform."[35] Hine's work is situated in a context of reform organizations and institutions that spoke a professional language and shared an optimistic perspective on the potential of reform to do good.

## THE HOME AS SWEATSHOP

The low wages paid for homework contributed significantly not only to families' use of children to maintain earnings but also to the intensity—the "sweating"—of the family's labor. Rheta Childe Dorr argued that the low wages meant that "to make anything at all the home worker must produce enormous quantities of work, she must toil inhumane hours, and all the children must toil with her."[36] Eleven-year-old James Gibbons of Roxbury and his sisters Helen, aged nine, and Mary, aged six, worked feverishly, sometimes until 11:00 p.m., repetitively tying tags to meet their desired quota (fig. 8.3). The most tags Helen ever tied in a day was five thousand, an accomplishment that earned her thirty cents. Her younger sister could tie only a thousand tags a day, however. In a "crowded" and "not very clean" household in Williamsburg, Massachusetts, the Weeks family strung wooden buttons, or button molds (fig. 8.4). Mrs. Weeks, her children (thirteen, eleven, and seven years old), and her grandchildren (ages seven, five, and four) worked continuously stringing button molds, "after school, holidays, etc." The most they ever earned, Mrs. Weeks told Hine, "was from $7 to $10" a month, "usually less," but "one time the children were all confined to the house by scarlet fever, and then she strung the most buttons she ever did."[37] In the Bureau of Statistics' report on the survey of Massachusetts homeworkers, Amy Hewes argued that "even the greatest industry and diligence can not raise the earnings above a level insufficient to maintain existence." Furthermore, she contended, "only in the rarest cases does home work bring in a living wage."[38] Dorr rightly asserted that, "as a cure for poverty, home work leaves much to be desired."[39] Photographs could not cure poverty either, but they could motivate philanthropic expressions of sympathy.

Investigators believed that in most instances homework provided supplemental income, rather than being a primary source of income. In the Massachusetts survey, married women made up almost three-fifths of the homeworkers over the age of fifteen, and 81 percent of these women homeworkers shared their household with an

Figure 8.3. "Home work on tags." (Hine 2980-A)

Figure 8.4. "Stringing wooden buttons (button moulds) in a crowded home." (Hine 3032)

adult male wage earner. Accordingly, "home work in Massachusetts is rather a side-issue, an occupation which may be taken up or dropped at will, and which supplements a regular wage from a factory worker." Most families would attempt to live off of their earnings from homework only as a "last resort." To escape poverty, some reformers argued, homeworkers needed only to take up employment in the factories. Others saw the low pay that homeworkers received merely as the product of an "abundant supply of labor."[40] This argument, however, failed to recognize the quandary that many mothers faced. Dorr, to her credit, acknowledged the possibility that "the abolition of home work [might] temporarily cause an increase in poverty," since "mothers of young children cannot go into the factories to work unless some means of disposing of their children is provided. The whole problem of the working mother waits for solution."[41] Homework was a flexible response to meet family needs, and those needs changed throughout the family's life cycle.[42] At certain stages, a mother might be able to leave her younger children with her older offspring and go to work in a factory for higher wages, while at other times working at home might be the only option available to her. A woman's work choices fluctuated as the family life cycle progressed.

Others looked at the relationship between homework and factory labor and emphasized the negative impact that homework had on the overall wage structure. Leaders of organized labor and reformers such as Mary Van Kleeck shared the belief that the low wages paid to homeworkers in general and to child laborers in particular had a tendency to drive down wages across the board. Unlike in other industries, there was very little talk about mothers' pensions when discussing homework, but they were an increasingly popular idea among reformers who sought to keep women from marching into the factories or from taking on more homework.[43]

An important way in which children contributed to homework was to serve as a go-between for the household and the factory, carrying raw materials and finished goods between the two. Rheta Childe Dorr wrote of how common it was in many industrial districts to see a girl "bearing on her head a huge bundle of unfinished cloth garments."[44] Hine made dozens of photographs of children toting large and small bundles, boxes, and packages between home and factory. Recall the photograph he took in Providence of a "tiny girl with big bag she is carrying home" (see fig. 7.7); the size of the bag over her shoulder, the younger child running behind her—these details help make the image both descriptive and evocative. Hine photographed an almost identical scene in Roxbury and used similar phrasing in his caption, describing his subject as "a little tot with a heavy load" (fig. 8.5). The boxes the young girl carries conceal half her face; the other half pokes out just enough for her to see what's ahead. The smaller girl behind her is running to keep up. We don't know their relationship but could guess that they are siblings. The image creates a sense of movement. The older girl means business and hurries along.

Figure 8.5. "Carrying tags. A little tot with a heavy load." (Hine 2967-A)

Hine provides a more common image of the garment worker in his photograph of Philip Descon (or Deacon), a young Boston boy who didn't know his age or even his "name very well," according to Hine (fig. 8.6). The photograph works in several ways. The boy balancing a large bundle on his head is a familiar representation of the garment worker. Hine's caption suggests a consequence of Philip's job as a garment worker: Philip barely knows who he is. Figures 8.5 and 8.6 capture the common sight that Dorr spoke of: the stream of children traversing the city with their heavy loads.

The journey between home and factory was a hard one, a fact that Hine expresses in his photograph of James Gibbons (fig. 8.7) and the "load of tags he is carrying a long distance from the factory on Vale Street to home." (Fig. 8.3 shows James working at home with his family.) The image of James is similar to the photograph of the "little tot with a heavy load"; both images are powerful and compelling.[45] In the case of James's photograph, the image is thickened with meaning because of what is going on behind James. The sidewalk heads off into the center of the horizon as James walks toward the camera. In the distance a dozen children are playing in the street, drawing a sharp contrast to James's solitary efforts. Children also stand along a rail just above James; they seem to be looking beyond him as they cheer on the boys at play.

Figure 8.6. "Philip Descon [Deacon?], 34 Charter Street." (Hine 2971-A)

In some instances a wagon owned by the factory brought goods to the home-worker's door, but more often children creatively assumed that responsibility.[46] In Somerville, Massachusetts, Hine photographed the wagon of a private contractor, O. H. Brown, who had created a business out of work frequently relegated to children. He owned four wagons in all and used them to transport materials from factories to people's homes.[47] Children, however, predominated in the transportation of goods between home and factory, using carts and carriages, gurneys, and whatever else they found at hand to ease their labors. Hine recorded some of the ingenious ways they had devised for ferrying goods back and forth. In his caption for a photograph of four young children in Roxbury, Hine reports, "They carry home the tags in all kinds of carts, baby carriages and in their arms. Often heavy loads and long distances."[48] A photo of another boy in Roxbury is simply labeled, "One way they get the tags home" (fig. 8.8). One of the strengths of this image is the woman in a white dress standing behind the boy. She has one arm on her hip and the other on the fence. She is a witness and perhaps the boy's protector. She certainly seems a pillar of the neighborhood, if not of society.

Figure 8.7. "James Gibbons and load of tags he is carrying a long distance." (Hine 2961-B)

Figure 8.8. "One way they get the tags home." (Hine 2965-A)

## HOMEWORK, EDUCATION, AND HEALTH

In many households, children's participation in homework superseded nearly all other activities. In Bridgeport, investigators spoke with a nine-year-old whose mother had died six months earlier. The young girl spent virtually all of her free time working at home with her aunt, operating a foot press. During the noon hour, she ran home from school to get an hour of work done. A brother and a cousin, both younger, helped out with the work. According to the report, "When she was asked, 'When do you have any time to play?' her answer was, 'Sometimes on Sunday.'" Often even sickness, rather than offering a reprieve from work, meant additional opportunities to labor. In Attleboro, Massachusetts, for example, Hine photographed Mrs. Gay and four of her children, ages five, seven, twelve, and thirteen, setting stones in "cheap" jewelry (fig. 8.9). Another son, ten years old, also worked but was not there when Hine took the photograph. The children regularly worked on the jewelry after school and into the evening, often past 8:00 p.m.; they didn't work much on Sundays and did housework (as opposed to homework) on Saturdays. Hine's caption notes that the children are all home with a sore throat but are nevertheless "working industriously."[49] Homework was typically the preeminent activity in the home, after sleeping.

Figure 8.9. "Mrs. Gay, 33 James Street, Attleboro, Mass., and children." (Hine 3143)

While investigators found that few children stayed out of school to engage in home-work, working in the home nevertheless had negative effects on children. Although the work was not dangerous or even physically taxing, the hours that children put in could be exhausting, and the physical and mental consequences cumulative. Reporting on the Massachusetts survey, Amy Hewes noted that "while home work does not directly interfere with school attendance," investigators found that "the child's strength of body and alertness of mind are impaired by long and late hours of mechanical, monotonous work." This finding was "supported by the testimony given by public school teachers in a town noted for the prevalence of home work," according to Hewes.[50] Children who were tired from their labors were less able to learn.

In many of the households in which the work of children took precedence over their schooling, the effects could be seen well into their adulthood. Investigators for the Rhode Island survey of the lace industry noted the high level of illiteracy among homeworkers. They reported that homeworkers who had begun working as children often continued in the same industry into adulthood, working both in the home and in the factory over the course of their lives. These investigators were not surprised that many foreign-born homeworkers had had no educational opportunities and could neither read nor write. They were more alarmed by the high illiteracy rate among native-born homeworkers who had begun their years of toil as children and bore the scars of that labor all their lives. "The majority of the illiterate were in the foreign-born group," they reported, but native-born home workers initiated into the world of work as children were also prevalent among the illiterate. They described one homeworker, "a woman of 50 years, born in Woonsocket, who had begun work in a textile mill at the age of 8. She had started, as most children did, by becoming a sweeper." Not only was this woman illiterate, but her "husband also could neither read nor write. He had started to work at 9 years as a sweeper in a mill." Another illiterate woman they inter-viewed, a native of Massachusetts, began working in a textile mill when she was only nine years old, bringing home a pitiful three dollars a month.[51]

Homework affected not only children's education but also their health. Reformers expressed considerable concern about potential health hazards, as they frequently found personal health compromised by homework. Hine makes frequent mention of illness in the captions for his photographs of homeworkers, suggesting, if not verify-ing, a connection between homework and ill health. In Worcester, Massachusetts, he photographed Mary George crocheting slippers with her twelve-year-old daughter, Elizabeth, and her thirteen-year-old son, Aaron. The children worked until as late as 10:30 p.m., the mother even later. Elizabeth, according to Hine, "has so much trouble with eyes that she is very much behind in school. Mother has eye trouble, too" (fig. 8.10). In his caption for the photograph of the South Framingham home of Rufine Morini (see fig. 8.1), Hine claims the children to be "anaemic." Children's health was

Figure 8.10. "Mrs. Mary George, 74 Southbridge Street." (Hine 3142)

greatly compromised by homework, in part because of the contagions that circulated with the goods the children worked on, broadening health risks.

Personal health problems translated readily into public health concerns. As Mary N. Winslow reported in the Bridgeport survey of homeworkers, public health was endangered when "members of the family are ill from contagious or infectious diseases." The danger was compounded by the "crowded conditions in the home of the typical home worker."[52] In Somerville, Massachusetts, Hine made several photographs of Jennie De Farsee (or De Farzen) at work crocheting underwear (fig. 8.11). Hine's caption for one of the photographs reports that Jennie had "an immense ring-worm on her face and another on her hand, but still she continued to work on the underwear." In his caption for another photograph that shows Jennie at work with her family, Hine refers to a "Home Work report about woman working here with running sore on limb."[53] Rheta Childe Dorr testified in her writing that she had "seen a woman in an advanced stage of tuberculosis, sitting up in an untidy bed, finishing coats. On the table beside her was a pile of labels" from a very reputable shop, and nearby a "boy, apparently tubercular, was sewing the labels in the coats."[54] In his caption for a photograph of a woman in an untidy kitchen in Leeds, Massachusetts, who is putting bristles

Figure 8.11. "Jennie De Farsee, 33 Horace Street." (Hine 2956-A)

into toothbrushes, Hine reports that although conditions in most of the homes he had visited were "pretty good," sometimes overcrowding and "filthy rooms are found and often tuberculosis with the work still going on."[55] Reformers found these spaces not only untidy and crowded but also undesirable and dangerous to the public welfare.

One of the greatest health hazards of homework was that it was rarely conducted in spaces set aside for such purposes, being most often carried on in the kitchen. Writing about the Massachusetts survey, Amy Hewes stated that "the kitchen is naturally the most convenient work place for the large numbers of workers who are married women; in their case home work alternates with housework at almost every hour of the day and they need to have their work close at hand where it can be picked up or dropped at any minute."[56] In their report on the Rhode Island study, Harriet Byrne and Bertha Blair noted that "the kitchen, which usually was also the dining room, in most households was the principal room, where the family congregated and where the work on the lace was done."[57] In Hine's photo of the South Framingham home of Desiderio Cella (fig. 8.12), Mrs. Cella and her children—ages thirteen, twelve, ten, eight, and seven—can be seen working on tags in a "dirty room with macaroni spread out on table being cut." Children tie tags at the same table, and Hine calls attention to the family's "washing also in dirty kitchen." Hine and other reformers were made uneasy

Figure 8.12. "Home of Desiderio Cella, 11 Coburn Street." (Hine 3137)

by the confluence of different and competing uses for the primary living spaces in the households of homeworkers.

The home was porous as far as homework was concerned. Sometimes homeworkers moved their work outside in search of fresh air, sunshine, a chat over the fence, or a chance to work with a neighbor and have someone over the age of twelve to talk with. Hine photographed many instances of homework being done outdoors, on household stoops, in back yards or side yards, and in city streets. "Sometimes," according to Hine, "they walk up and down the street as they work."[58] It was a "typical sight," Hine said, to see "family and neighbors working on tags" on the steps (fig. 8.13).

Where they could, public health officials and reformers attempted to regulate domestic space and homework conditions. Writing about homeworkers in Massachusetts, Hine argued that although "the people are supposed to do the work only under certain restrictions," when the "inspector and the one who delivers the goods are not around, they do as they please." Hine photographed Annie Fedele of Somerville crocheting underwear in a "dirty kitchen" (fig. 8.14), as she stood out on the street, and sat in her "filthy back yard" (fig. 8.15). Annie and her family basically ignored all regulations. Elsewhere in Somerville, homeworkers crocheted underwear

Figure 8.13. "A typical sight. Family and neighbors working on tags on door steps." (Hine 2986-A)

under similar conditions, with "mother and children and surroundings all filthy." For Hine, this provided a "good illustration of the difficulty in trying to regulate Home Work."[59] The decentralized and semiprivate nature of the work made any enforcement of regulations extremely challenging.

In Massachusetts, which was one of the few states with a licensing system, enforcement was uneven at best. Licenses were required for all Massachusetts families who worked on wearing apparel in their homes, but investigators for the Massachusetts study found that of 214 families engaged in the home manufacture of apparel in twelve municipalities outside of metropolitan Boston—Chicopee, Foxborough, Haverhill, Leominster, Newburyport, Northampton, Reading, Salem, Framingham, Springfield, West Springfield, and Worcester—only one (a corset worker in Worcester) had a license. And in only one Massachusetts household did investigators find that a license for homework had been revoked, because the homeworker "had persisted in working on articles of wearing apparel in her kitchen." For the most part, authorities exercised no "control whatsoever in regard to conditions under which [homeworkers] carried on their work."[60]

The conformity of homeworkers' working conditions to the standards established to govern child labor very often escaped public scrutiny and legal sanction. The Massachusetts Board of Labor and Industries bemoaned the fact that, under the

Figure 8.14. "Annie Fedele, 22 Horace Street." (Hine 2951-A)

Figure 8.15. "Annie Fedele crocheting on underwear out in the filthy back yard." (Hine 2952-A)

then-current system of homework, "if there is violation of the child labor law," the manufacturer "cannot be prosecuted for the same." Massachusetts law provided that "if any child or woman shall be employed in more than one such place [of manufacture], the total number of hours of such employment shall not exceed fifty-four hours in one week." However, the board noted, "it is a common practice in some industries for girls to take home from the factory where they are employed work to be done at night after the day's task is ended. Thus they continue the employment of the day with no relaxation such as might come from a change of work. This is but another evidence of the different standards applied by the labor laws to work done in a factory and the manufacture of the same product at home." Homeworkers needed the same protections as factory workers, but the state had thus far exhibited a laissez-faire attitude toward homework.[61] Hine's photographs and the New England surveys of industrial homework pushed the public to demand greater protections for children working at home.

Owen R. Lovejoy, general secretary of the National Child Labor Committee, shared the belief that "the employment of children in tenement homes is one of the most subtle forms of child labor and one of the most serious in its effects upon society." The problem was that "inspection of private homes is always difficult. . . . Despite the utmost vigilance, the employment of children from ten to five years of age for excessive hours, under the direction of their own parents, is not uncommon in communities that would not tolerate the employment of children in factories or stores under fourteen years of age." This "evil" existed widely in northern U.S. cities, where it threatened children's education and development as well as the public's health. In addition, "legitimate" businesses were hard pressed to compete with manufacturers who could "economize charges for light, heat, rent and mechanical power by dividing these burdens among the helpless tenement dwellers and producing cheap goods at the sacrifice of decent wages."[62] Without community consensus and some form of self-enforcement, it would be nearly impossible to halt the use of cheap labor to make cheap products and the subsidization of companies by allowing them to transfer a share of their overhead expenses to their workers.

Some believed that as long as homework persisted, there could be little success in protecting children (fig. 8.16). In its report on the Bridgeport survey of homeworkers, the U.S. Women's Bureau stated flatly that "home work should be abolished."[63] The bureau acknowledged that the abolition of homework would put a particular burden on "underpaid women" whose income was intimately connected with their children's labor. How to abolish homework and at the same time "avoid the privations" that those women could be expected to suffer would have to be "decided by the community." Well-meaning employers had recommended the "regulation of home conditions through a system of inspection." The discussion, however, was too often "left

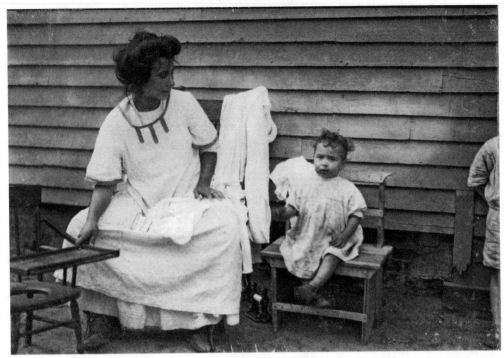

Figure 8.16. "This is typical of the home work on underwear in these quarters." (Hine 2977-B)

unfinished of the more fundamental recommendations, such as possible arrangement of short shifts for women who could not be at the factory all day, payment of equal rates to workers at home and in the factory, and possible adjustments which could be made by social agencies and city authorities in families where there is more need of social treatment than of home work."[64] Homework was a community question, and beyond the reformers themselves, it was not yet clear how many in the community were inclined to eliminate the practice altogether.

In photographing homeworkers, Hine uncovered low wages and long hours, running noses and running sores, dangers of disease, irregularity, and excessive labor. Hine found what he was looking for, but neither he nor any of the surveys of the period gave much, if any, attention to what went on at the tables where mothers and children worked together. We know that neighbors met in yards and alleys to work and share one another's company. What did they talk about over those long hours? Did they laugh? Did they sing, tell stories or jokes, read or recite, pass on knowledge of the family past? Did they dread drunken husbands and fathers coming home? Were they content or discontent, angry or accepting of their lot in life? Did it matter that homeworkers were largely portrayed as non-natives? At least the foreign-born could be excused for violating middle-class values they simply hadn't learned yet. But the prevalence of native-born homeworkers suggests that they may not have accepted the middle-class valuation of the separation of home and work. Perhaps working-class people valued the supplemental income from homework in ways that middle-class reformers had difficulty understanding. Workers were aware of the vagaries of existence and the need to have more than one source of income. All of the studies showed that the income taken in from homework did not make much difference in these households; that may well have been true, but did the income have a psychological value, making homework more difficult to pathologize among certain groups?[65]

Because this child labor occurred in the home and under parental supervision, many wanted simply to ignore the issue. Hine characterized the task of protecting children from the evils of homework as "a job for Hercules" and said that he could "see that worthy gentleman approached by the king or some other of the apologists of the day, and putting soothing hands on his sturdy shoulders, they might have said, 'Now look-a-here, Mr. Hercules,' (or did they call him Doctor or Professor?), 'them-there stables haint so bad. We been livin' with 'em for years an' we're purty healthy.'" But Hine believed that "what we must do" is to "turn the River of Public Sentiment into them and clean out the whole business." Through his photographs he hoped to state emphatically, "The home is for the family—Industry Keep Out."[66]

CHAPTER 9

# Working-Class Communities

———

IN THE MASSACHUSETTS CITIES OF Fall River, Lowell, Lawrence, New Bedford, and North Adams, child labor was not a reform movement but an integral part of working-class life. When Hine photographed a sign on the Small Brothers mill soliciting the labor of boys and girls (fig. 9.1), he accompanied his image with the notice that this Fall River factory had a "reputation of employing mostly children." The sign was not a casual appeal reflecting a momentary aberration in the social order: it was permanently attached to the factory, signaling the fixity of child labor in working-class life. Lewis W. Hine's photographs, taken between 1909 and 1917, documented the ubiquity of child labor in New England working-class life. They were produced for the consumption of Progressive reformers and reflected their concerns and their critique of working-class life. Each photograph in Hine's intensive survey of Massachusetts textile centers contributes a small fragment to a portrait of the making of the American working class. But some details are missing from this portrait. Hine was a reformer among those to be reformed, and his collective portrait is absent depictions of critical aspects of working-class life, most notably images of working-class institutions and expressions of working-class consciousness. Hine was trying to portray children as helpless and worthy of social welfare reforms, rather than as active agents in their own lives.

## CLASS AND COMMUNITY

Historians have written extensively and eloquently on the formation of class and community in New England, particularly in its textile centers. An abundance of community studies delineate the contours and boundaries that shaped working-class culture.

Figure 9.1. "Boy & Girl Wanted. Printed permanent sign." (Hine 4207)

Roots in artisanal, agricultural, or commercial culture, patterns of paternalistic cor-poratism, cultural traditions of religion or trade unionism, ethnic composition, and urban disposition—all shaped the formation and expression of class.[1] Hine photo-graphed the working-class youth of New England in towns and cities throughout the region. In some cities, such as Lowell, Fall River, New Bedford, and Lawrence, Hine photographed intensively, working in multiple neighborhoods and among diverse ethnic groups, each group having a unique history but sharing distinct characteristics of class with the other groups.

Class is a relationship not easily captured on film, though Hine left behind hundreds of fragments that can be sewn together to craft a collective portrait of New England labor. In what ways did Hine photograph the making of the working class? Did his photographs capture a working-class style, a dialect, a working-class consciousness expressed in gesture, posture, or poise? Hine framed his images, composing narratives to meet reformers' needs, while the children who were the subjects of Hine's photo-graphs came as they were in their daily lives. Those lives were expressed in the dirt on the children's faces, the too-large shoes, the too-small coats, the well-worn clothes handed down between siblings, the newspaper-pattern dresses hand-sewn and worn with a style of their own. Their daily lives were also expressed in a smirk, in a smile, in a flexing of muscles, in an arm thrown over the back of a friend or in elbows knotted tightly as young women walked down the street and protected themselves from the gauntlet they had to run on their way to work. These were passive expressions of class. Hine's camera could not capture a worker's whispered advice to "walk, don't run" on the job or "avoid speeding up when the boss is around." Hine's photographs, in part, defined the parameters of class narrowly and passively, each image vested with the values and virtues of middle-class Progressive faith and composed of working-class subjects. But certain types of scenes from working-class life were not present in Hine's photographs: there were no institutions of working-class life—no union halls, mutual benefit societies, or ethnic ballrooms—and no collective activities organized around class or ethnicity—that is, no festivals, parades, or political activities. Expressions of class consciousness or self-organization were absent from Hine's photographs.

The role of children and their contributions to working-class survival have been acknowledged as considerable by scholars such as John T. Cumbler and William F. Hartford. Low wages in the textile industry put pressure on workers to embrace a family economic unit rather than depend on a single breadwinner. Writing of workers in Fall River, Cumbler notes, "Even the more successful textile workers had to withdraw children from school and place them in the mills." For unskilled textile workers the circumstances were more precarious, and "even with children at work life was a much greater struggle." With several children at work, a textile worker's family could enjoy a modest standard of living, "but conditions were mean indeed," according to Cumbler, "for those with offspring too small or too sickly to enter the mill." Working children contributed more than 33 percent of family income. Thus, among the working class, a large family could be a collective asset rather than a financial drain.[2]

The children came to Hine's camera as they were, without staging or an agenda. The willingness of the children and their families to expose themselves to an outsider, to be photographed for display, was reflective of the naturalness of child labor. To them, child labor was not an aberration but a common fact of life made necessary by low wages in the textile industry. Unlike their professional counterparts, working-class people did not necessarily see child labor as a comment on their parenting abilities. From their viewpoint, child labor was not a threat to citizenship or a moral danger to the nation. Child labor could be a key to family survival, or it might simply provide a young boy or girl access to commercial entertainments and a measure of personal freedom. It might also be the only evident path to future employment, and clearly many children preferred work to school. Exploitative and often exhausting, but hardly unusual, child labor was everywhere. An example of the clash between professional and working-class values came when Hine went looking for evidence of social wrong

Figure 9.2. "While I was photographing these workers (Berkshire Mills) the watchman dragged out the smallest boy." (Hine 2347)

at the Berkshire Mills in Adams, Massachusetts, and was met by the watchman, who didn't attempt to hide the children working in the mills. To the contrary, "the watchman dragged out the smallest boy," according to Hine, "saying, 'Here, photograph 'Peewee'" (fig. 9.2). There was no shame, no expression of embarrassment or guilt among the watchmen and workers lounging about during their lunch break. While Hine no doubt intended for his photograph to make a statement on a transgression of societal norms, the watchman illustrated the normative nature of child labor in working-class communities in New England.

Children laboring didn't transgress or violate the cultural values of the working class in the ways that it appeared to contradict the cultural expectations of middle-class reformers. John Parent, the fourteen-year-old subject of a photograph Hine took outside a mill in Salem (fig. 9.3), was very much part of the working world, as he leaned against the wall at the factory gates and watched workers pass by to enter or exit. Although he is presented in Hine's photograph merely as a child who works, John was almost certainly also a brother, son, neighbor, friend, and schoolmate to others. He carried those connections from the community into the factory. The three women

FIGURE 9.3. "Boy is John Parent, 14 Congress St." (Hine 2631)

in the photograph with John could well be his mother or aunt, his sisters, or his neighbors. The key point is that child laborers entered the factory with full identities, linking them intimately with their fellow workers.

Child laborers often saw themselves in a more favorable light than did reformers, not in spite of their identity as workers but because of it. Some of the child laborers Hine photographed displayed great pride in being a part of the industrial workforce. This pride was also evident among their parents and neighbors' parents, as can be seen with the Lancashire weavers in Fall River.[3] A boy who appears in Hine's photograph of underage workers at the Eclipse Mills in North Adams, Massachusetts (fig. 9.4), poses with perceptible pride; the pipe in his mouth wafting smoke, his thin arms flexing to show off his underdeveloped muscles, his back straight and his posture upright, the young mill worker takes pleasure in his "manly" bearing. As children entered the industrial workforce, they were also transitioning from childhood to adulthood. Like their parents, neighbors, and friends, they became workers and joined the industrial proletariat.

The children didn't see themselves as victims. That working conditions were exploitative was a part of daily life for all workers, irrespective of their age. Children were cheap labor, just as their mothers were cheap labor. They aligned themselves with their

FIGURE 9.4. "Arthur Chalifoux, 3 Rand St." (Hine 2339)

coworkers in what was clearly the most important step in the development of their class standing. Although this was true for girls as well as boys, boys tended to mix themselves into large groups of their cohorts. Young girls did not gather into large clusters in the same way. Their images outside of work tended to link them together in small groups or situate them not among their peers but with their families.

Working in Massachusetts cities under the direction of the Massachusetts Child Labor Committee gave Lewis Hine the opportunity to photograph children in multiple contexts. Hine extended the visualization of his subjects by following them out of the workplace and into their communities and homes. When Hine photographed boys, he often placed them in large gaggles. He took a different approach in creating his portraits of young girls. He created complementary photographs that linked the girls less to each other than to their families. At the same time, he rarely photographed a boy both in a work context and in a domestic one. Consider, for example, Hine's photograph of Henry Fournier (or Fourner), a sweeper and cleaner in the #2 spinning room in a mill in Salem (fig. 9.5). It is one of the more effective photographs of a young boy in a textile factory in the North. Hine places Fournier slightly off-center in the image; two long spinning frames fade into the distance behind him. Hine composed the image so that it would accentuate the diminutive stature of Fournier and

FIGURE 9.5. "Henry Fourner [Fournier?], 261 Jefferson St., Castle Hill." (Hine 2632)

highlight the inappropriateness of his being in such a place at his tender age. Hine's caption merely gives Fournier's name and address and identifies him as a sweeper and cleaner. When the Massachusetts Child Labor Committee used the image for its annual report, it altered Hine's captioning to play more directly to reformers' concerns: "Dangerous playthings for little children. This boy has been at work only two months and has not yet lost his childish beauty." The committee's caption for a comparable photograph appealed to the fierce pride of locals: "This is not South Carolina—it is Massachusetts."[4] Photographing Fournier outside the factory, Hine lined him up with his contemporaries and in the accompanying caption identifies Fournier as the smallest boy in the image.[5] In spite of his size, however, Fournier was still part of the group and drew his identity from it.

Hine's photographs of young girls working in the mills of Winchendon, Massachusetts, reflect gendered contexts. Thirteen-year-old Mamie La Barge (fig. 9.6) was Hine's most photographed female subject, her image having been captured by Hine at least fifteen times, either individually or as part of a group. Hine obviously must have felt that he had hit pay dirt with this beautiful young mill worker. Among the images he made of Mamie is this photograph of her standing alongside her machine, a framing

FIGURE 9.6. "Mamie La Barge at her Machine." (Hine 2365)

that he used countless times to call attention to the incongruity in size between a young worker and the machines he or she tended. In this photograph, Hine has given Mamie both physical and cultural props. He ties Mamie to her machine with a long strand of thread, while his caption notes that she is "under legal age." Hine also produced several photographs of Adrienne Pagnette, including an image of her in front of a spinning frame. In his caption for the image, not only does he condemn Adrienne for being "illiterate," but he also burdens her with the judgment that "Stooping, reaching and pushing heavy boxes is bad for young girls adolescent."[6] Hine's message was simple: the factory was no place for young women. But that ideal did not reflect life in the region's textile centers.

## FALL RIVER, MASSACHUSETTS

No U.S. city had as many children laboring in the cotton textile industry as Fall River did, relative to its size. According to the U.S. Census of Manufactures for 1905, Fall River contained nearly one-seventh of all the cotton spindles in the United States and more than twice the number of cotton spindles that could be found in any other city in the country. By 1905 effective state regulation had reduced the proportion of cotton factory workers who were under sixteen years of age to 6.3 percent in Massachusetts, and in the Fall River factories that proportion had shrunk to just 5 percent. Throughout the country as a whole, children represented 12.8 percent of all textile operatives; in some southern states that number rose to as high as 22.9 percent.

Despite the relatively low ratio of children employed in the cotton mills to the whole number of textile operatives in Fall River, the size of the industry was so great that there likely were more children employed in Fall River than in any other city of its size in the country, according to the U.S. Bureau of Labor's *Report on the Condition of Woman and Child Wage-Earners.* That report also noted that the number of age and schooling certificates issued for fourteen- and fifteen-year-old children was almost certainly higher in Fall River than anywhere else. (State law required a job applicant between the ages of fourteen and sixteen to present his or her prospective employer with an age and schooling certificate, which certified that the holder had attained a minimum education in the judgment of the local school authorities.) Like other textile centers in New England, Fall River was an ethnically diverse place, with "twenty-two races of people" making up the city's population (fig. 9.7). The children employed in the city's cotton mills were born in eighteen different countries.[7]

The authors of the labor bureau's *Report* maintained that Fall River, "if unfettered by State regulation, would offer more opportunities for the employing of children than any industry in any other locality of the country." Fall River was a working-class city in which the supply of young workers was certainly plentiful. Massachusetts, however, had been among the first states to regulate the conditions for child laborers and possessed "advanced and efficiently enforced child-labor laws" that balanced the "industrial tendencies of the factory system and the economic demands of the laborer's family . . . against the restraints of state regulation."[8]

A particular gesture, one of friendship and solidarity, surfaces in many of Hine's photographs, in Fall River and elsewhere. Hine repeatedly photographed the physical affection between boys, the draping of an arm around a comrade's neck. He produced many images of boys casually leaning against one another and expressing their friendship with ease. While the sight of young women holding hands as they walked the city streets was fairly common, a boy's physical expression of friendship was limited to hanging his arm over the shoulder of his friend. Solidarity in these images involved personal friendships rather than social or political commitments. Hine made no photographs of collective activity among working-class youth.

One of Hine's most common photographic compositions was the frontal image of a large group. Hine took this type of photograph throughout his tenure with the National Child Labor Committee. When Hine photographed a large group of boys in front of the Indian Manufacturing Company in Indian Orchard, Massachusetts—noting in his

FIGURE 9.7. "Portuguese. Spinner." (Hine 4246)

FIGURE 9.8. "Group in front of Indian Mfg. Co. Everyone in photo was working." (Hine 2361)

caption that all of the boys worked at the factory—he let his subjects arrange themselves as they pleased (fig. 9.8). In this image, as in many like it that Hine captured,
the boys pose for one another as much as for the photographer: cigarettes and pipes
extend from the corners of their mouths; hats are tilted; arms are crossed in front for
a manly standing, while other boys stand with a hand on their hip. The boys have put
on a show for Hine—they are constrained, to be sure, but it is they who have framed
their cool countenance. The children have gathered collectively but are not collectively
engaged. This is an image of boys as chums, not as comrades.

No matter the working-class community he worked in, Hine captured one common
quality, the physicality of the boys with one another. It was not unusual to see the boys
pile onto each other. On occasion, Hine threw himself into the center of a crowd of boys
as they crammed together to make their way into the photographer's frame. In taking a
noontime photograph of boys who worked at the Eclipse Mills in North Adams, Hine
found himself in one such scrum (fig. 9.9). He focused on the boys closest to the camera
so that, as the depth of perception grows, the image becomes out of focus and more
impressionistic (though it falls far short of the work of the early twentieth-century
photo-secessionists, who valued artistic expression over objective clarity). The boys are
highly intimate with one another, but their affection is indiscriminate. One boy seems

to rest his face on another boy's head as he giggles and tries to hang on to an open space between himself and the photographer. Another boy pokes his head through to show his smile to Hine. We cannot tell how verbal they were in the expression of friendship, but they hesitated not one bit to grab hold of each other and squeeze into the frame. Physical contact and affection were important in the boys' lives.

One beat in the rhythm of the walk to work was the verbal assault that women faced daily. Hine often photographed women as they walked past groups of men leaning against a wall or clustering about the factory gates, passing judgment on them. These were men that the women had to pass every morning on their way to work. The time before work was when the young women were most vulnerable. As male and female workers arrived at the factory and awaited the opening of the gates, they often milled about. Undoubtedly, many had their own routines—meet up with a friend, chat about the latest news or sports, gossip, gamble, and so on. This was a period of the day in which men and women, boys and girls, all met indiscriminately, mixing and dividing themselves as propriety and local custom permitted. Young men often used the time to yell sexually harassing comments at young women. In New Bedford, Massachusetts, Hine photographed a group of young girls "running [a] gauntlet of men making remarks (sometimes vile)" (fig. 9.10). This was further evidence, Hine

FIGURE 9.9. "A group of mill workers. Albert Duquette on top." (Hine 2344)

FIGURE 9.10. "[Young girls running gauntlet of men making remarks (sometimes vile)]." (Hine 2798)

believed, that young women didn't belong in the workplace; they belonged at home, safe from such verbal assaults.

Hine captured this scene over and over, an indication of the widespread nature of sexism in working-class life. In these moments of exchange as they were on their way to work, boys and girls learned their roles. The young women generally faced the inter-generational gauntlet on their own, without the support of older women who might know the best way to respond. Young boys, on the other hand, learned at their father's knee, so to speak. Adoption of sexist behavior was part of the passage into manhood in working-class lives. Women were not entirely passive by any means, although one can clearly see in Hine's images those girls who were brave enough to respond and those who were more likely to blush and move quickly past the danger. Hine's caption for his photograph of a group of adolescent girls and young men in Chicopee, Massachusetts, indicates his understanding of the educational value of such moments (fig. 9.11): he observes that "a young fellow made a very vulgar remark and they all laughed. Next day I heard one of the girls use profanity before all the men and boys." It was precisely this type of education that most concerned reformers interested in promoting chil-dren's moral development.

Although they are much rarer in his archive, Hine took many photographs of

FIGURE 9.11. "[Group of girls, adolescents. A young fellow made a very vulgar remark . . .]." (Hine 2656)

children walking to and from school rather than to work. For instance, Hine made a series of photographs of children between the ages of fourteen and sixteen on their way to the Watson School in Fall River (fig. 9.12). The schoolchildren, with their stylish hats and bright, spotless dresses worn with an eye toward fashion, stand in marked contrast to their laboring counterparts, whom Hine also photographed, at work and at play.[9] The young workers, Hine captioned, were employees of Kerr Thread in Fall River and posed for Hine at the noon hour. The boys were nearly all between fourteen and sixteen years old, and they too had an informal uniform of sorts—a cap worn at a tilt, overalls, and rolled-up shirtsleeves.

Hine's photographs depict the changing orientation of the child labor reform movement, particularly in places like Massachusetts, which had the most advanced child labor laws in the country. Reformers began to move well beyond simply worrying about minimum age requirements. For instance, Hine photographed a baseball game between young male workers at the Kerr Thread Mill that was the product of a new initiative by the State Board of Labor and Industries, working in conjunction with the Massachusetts Child Labor Committee, to "get communities to build up the health and strength of young workers."[10] Child laborers, reformers now argued, had

FIGURE 9.12. "Going to School. About 14–16 years." (Hine 4216)

three fundamental needs. A "constructive" agenda on child labor in Massachusetts would have to include more education for young workers, more recreation to keep them from "wasting away in the mill," and better medical care to determine the overall health and condition of each child. Recreational activities and medical attention were relatively easy to provide. Since so many children saw little need to continue with their education past grade school, educators and other reformers increasingly emphasized the need for the development of some form of alternative industrial or vocational schooling.[11]

Hine photographed the rudiments of the corporate welfare system taking shape, and child laborers, like all workers, benefited from the resulting improvements. The Kerr Thread Company, like the Amoskeag Manufacturing Company in Manchester, New Hampshire, embraced the new reforms aimed at improving conditions in the mill in order to improve relations between labor and management. The company could build loyal workers and healthier citizens by attending to their needs and providing protections for their well-being. And better that the children play baseball than listen to the appeals of labor agitators. Hine photographed a group of female Kerr Thread Mill employees, all over the age of sixteen, who are looking down on him and "having fun with the camera man" (fig. 9.13). Hine's caption reports that conditions in the mill

FIGURE 9.13. "Kerr Thread Mill. All over 16. Having fun with camera man." (Hine 4208)

were very good. Hine specifically notes the caps the women are wearing "to protect hair from dust and to keep hair from getting tangled in machinery." The history of the textile industry was full of instances in which workers had gotten clothing or hair caught in a machine or belting. For instance, there had been the dramatic testimony of fourteen-year-old Camella Teoli, submitted to Congress during its investigation of the 1912 strike at Lawrence, Massachusetts, in which she described how "the machine pulled [my] scalp off."[12] Teoli's testimony was a cautionary tale that connected work-place injury with militant syndicalism while also serving as an example of what could happen when adolescent girls entered the textile mill.

Hine photographed other reforms at the Kerr Thread Mill, such as the creation of a small hospital / rest area where workers could receive immediate attention from a doctor or a seventeen-year-old "practical" nurse (fig. 9.14). Mr. Kerr, the owner of the mill, was sincerely interested in the well-being of his workers, and never more so than when their interests coincided with his own. Kerr was proud of the establish-ment of a medical station, and Hine captions that "Mr. Kerr thought it a fine thing for girls and for mill." The demand for the service was steady; as many as fifteen to twenty employees sought help on certain mornings. Here was an excellent example

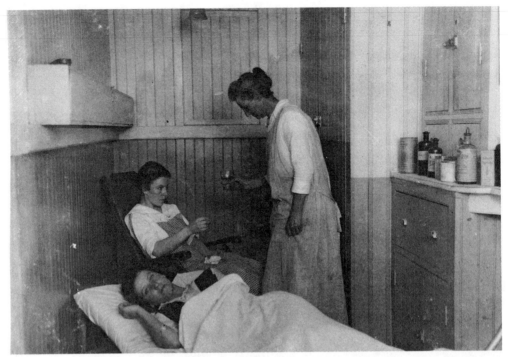

FIGURE 9.14. "1st Aid—Kerr Thread. Rest room—one step beyond the emergency box." (Hine 4206)

of divergent yet reinforcing benefits derived from improving conditions in the mill. The young women received medical assistance they probably would not get otherwise, and the company benefited by "saving days off." For the workers, their identification with the company grew as they felt its protective arm reach out to attend to their basic needs. The company was learning that the stability gained from worker satisfaction made such reforms a profitable investment.

Another lesson to be taken from Hine's photographs at the Kerr Thread Mill is that there were limits to the reach of company policies. Working-class culture could not be overridden or easily overturned, though it might be ignored by Hine. Children learned by observing the adults around them. One of Hine's photographs shows a group of men at the top of the stairs at the entrance to the Kerr Thread factory. They are laughing and hanging on to one another. In front of them stands a young boy; Hine captions that the boy "always gets jollied" by the older men around him. Sadly, the boy returns the jollification when "he answers the men back as tough as you please." Obviously, there was mutual affection between the little boy and his coworkers, though the boy's father was not among those pictured; he was at a nearby saloon, where the boy "went in to get money." Efforts at providing a more uplifting workplace were sometimes countered by baser influences, and few were more powerful than alcohol.

Material want was widespread in working-class communities, and child laborers, like their parents, contributed to their family's survival by scouring the city for anything that would defray an expense or meet a family need. Scavenging was widespread among working-class children. It was casual labor, irregular in every way, impossible to measure precisely, and deemed by many to be a major problem. A Massachusetts Child Labor Committee study found that of 2,128 children in one Boston school district, 125 were "regularly engaged in picking over dumps of ashes, garbage and other refuse with their fingers (seven children less than 13 years old in one family), 500 went for ice, 117 to the market for produce, 83 for fish and 61 for potatoes, while 867 gathered wood, 225 of them selling it outside of their families."[13] Investigators for the committee estimated that in Boston alone, more than 500 children picked over the dumps. In its report on the study, the committee told of a typical dump in which "fifty-five children from five to fourteen years old were moving slowly over an acre of rubbish, bending their short backs so that their heads were almost touching the refuse." As the children worked, they "stirred up dust" as well as the dump's "vile" odor. A dozen men picked through the dump alongside the children, loading their wagons with junk while the children filled their handcarts. The committee urged the public and the authorities to "keep in mind the health of these children" and called for the passage and enforcement of a state law or city ordinance banning children from the dumps.[14] Nothing came of the call; the dumps remained the picking ground for young laborers.

Hine photographed children working the dumps in several New England cities, but none of the other images possesses the depth of caption or matches the disturbing story of Oscar Revinsky (fig. 9.15). Revinsky was well known to local authorities in Fall River. The Society for the Protection and Care of Children opened a case on Revinsky in 1910 and for the next five years observed his steady disengagement from reality. By 1912, Oscar was neither working nor attending school. He had once attended a "baby grade" but was expelled. He was considered to be mentally deficient, but the community offered few resources to help with the care of the mentally ill. Oscar's parents at first rejected the idea of committing their son to a mental health facility in Wrentham, Massachusetts; like many parents, they didn't feel comfortable surrendering their child to an unknown caregiver. By 1916, however, Oscar's father had changed his mind and sought to have his son committed, as he was unable to care for him. Oscar spent all his time in dumps and never washed. He no longer came home for meals but instead ate "from [the] dump" and stole "from dinner pails." Hine noted that Oscar's neck was "covered with scars and boils." At the time Hine photographed him, Oscar's mental state was confused; Hine's caption notes that Oscar wasn't sure where he lived, and when Oscar was asked whether his father was alive, he replied, "No, he's a milkman." Hine's photograph draws Oscar's problems starkly, making it easy to extrapolate the cost of scavenging to the larger community

FIGURE 9.15. "Oscar Revinsky(?) — Born Jan. 11, 1900, 15 years old." (Hine 4322)

and prompting the viewer to consider what role the state should play in addressing Oscar's particular needs.

One of the most common forms of scavenging, particularly in Boston, was gathering wood. The Massachusetts Child Labor Committee reported that "small boys, and even girls, permeated to the skin with dirt, their ragged clothes torn by nails, pile up great loads, heavy enough for two union laborers to carry, and drag or struggle under them for a mile or two." The committee also told of an eleven-year-old boy living in Boston's North End who "had worked an hour at Copley Square filling his cart, building the sides up with sticks, until the load of timber and boards was about four feet high." With great effort and exertion, "his face flushed and strained," the boy and his nine-year-old brother began to pull the overloaded cart forward, stopping frequently as they struggled to get their load home. The eleven-year-old typically earned "the exceptionally large sum of six dollars a week." The committee also cited a paper by Dr. Joel E. Goldthwait in which he reported that children had been "permanently injured by the constant strain of carrying loads continuously for long distances."[15]

Scavenging tempted children to steal as well as to gather. The committee observed that "in all scavenger work there is a strong incentive to theft. It is not the doing of useful chores, it is the regular and systematic work of scavengers in industry, and it leads to crime." Boys who gathered barrels to scavenge the wood frequently resorted to stealing them, while children in search of coal sometimes jumped on passing trains "in order to steal coal." Commenting on the "gathering of refuse from markets, ash barrels, and freight yards," the committee argued that the practice was unhealthful, brought "small advantage to the family," and involved "so much petty larceny" that, like children's work in the dumps, "it ought to be stopped." The committee concluded, however, that wholesale prohibition of the gathering of wood and coal was unfeasible, and so it recommended "regulation by license" as the best remedy.[16] Certainly it was not difficult to spot children gathering coal and wood (fig. 9.16), but regulating them was another matter altogether. As for the arguable benefits of gathering refuse, scavengers and their parents probably understood the matter of "small advantage" differently from reformers, given the tight budgets many working-class families lived on. Their side of that equation might have been something like, why buy coal when you can get it for free?

FIGURE 9.16. "[Stealing coal from railroad coal-yard.]" (Hine 4718)

Working-class attitudes toward theft allowed greater latitude for interpretation, particularly if the theft was driven by necessity. Stealing to supplement family resources was common, and children were quite often the ones doing the taking. Working-class individuals made their own differentiation between stealing from neighbors or local peddlers and storekeepers and stealing from the large, impersonal railroad corporation. The bottom line was that poverty and want could help to rationalize many things. In Florence Kelley Wischnewetzky's translation of Friedrich Engels's *The Condition of the Working-Class in England in 1844,* Engels argues that "want leaves the working-man the choice between starving slowly, killing himself speedily, or taking what he needs where he finds it—in plain English, stealing. And there is no cause for surprise that most of them prefer stealing to starvation and suicide." There was not widespread acceptance of stealing in working-class communities, just a different understanding of what the crime of stealing entailed.[17] Children were filling in the gaps in family resources. They were adding their bit. Probably no one in a child's family asked too many questions about where the child's contribution came from; all were just grateful it was there. At the same time, there was likely little tolerance for stealing from one's own kind.

Reformers maintained that exercise brings growth—not simply physical growth but mental growth as well. Many, like William J. Lee, recreation supervisor for the New York City Parks Department, saw work as a higher form of play. The teamwork that children learned on the handball court or baseball field, he argued, formed the basis for building and engaging in teamwork for the rest of their lives. Given the increasing importance of play in reformers' eyes, failing to steward such an important element of children's development was an unwise choice. A new feature in the world of play was the growing embrace of play equipment—swings, slides, monkey bars, and jungle gyms—to facilitate appropriate forms of play for the longer-term development of the child. Hine photographed a group of working boys playing on playground equipment at Sandy Beach on their day off (fig. 9.17). They are climbing and hanging and swinging, building their muscles, which reformers thought vital to children's physical development. They are actively engaged in creating play of their own, which could benefit their mental development.

The continuing growth of commercial entertainment was one of the primary inducements for children to work. Children were among the most avid consumers of commercial recreation and amusements. Unlike noncommercial play in the streets or on playgrounds, commercial recreations depended on the market for survival. They were businesses that sought to exploit the public's desire for fun, excitement, and novelty. They marketed themselves to young children, encouraging their participation. To join in such pleasures, however, a child needed money. Children learned early on that to obtain personal pleasures, work was the key ingredient, as it was in simple survival.

Commercial entertainments like vaudeville endangered children by not providing

FIGURE 9.17. "On the beach. Sandy Beach." (Hine 4180)

supervision of their play and putting them into promiscuous positions. On a Saturday evening in Fall River, Hine photographed a "long line of boys and men" at a vaudeville show" (fig. 9.18). The show appears to have had appeal across the ages, and the women standing at the head of the line suggest that, despite Hine's captioning, neither the show nor the venue is all-male. One of the dangers posed by commercial recreation was that young men and women were often left alone to enjoy the entertainment without the requisite supervision.

In an image that he created at Sandy Beach, Hine portrays another busy scene (fig. 9.19). Hine's caption for this photograph notes that a "penny picture machine" is "attracting crowds," but most of the people funnel into the building with the sign announcing that there's "Dancing" inside. Hine calls the viewer's attention to the two girls in the photograph poised before the dance hall, stalled in talk, neither entering nor walking away. While all might appear fine at first glance, Hine challenges the viewer to look more deeply and, in light of the additional information that Mr. Tebbutt offered, to understand that conditions in the dance hall were "bad."[18] Yet the poor conditions inside did not seem to deter the crowd from entering, nor did they stop "Saturday evening dancing" from being "about the same as this."

A common experience in working-class life that Hine did not photograph was

FIGURE 9.18. "Long line of boys and men at Vaudeville Show." (Hine 2769)

industrial conflict. In 1904 and 1905 there were 180 strikes in the textile industry in Massachusetts. The "great textile strike" of those years began in Fall River on July 4, 1904, when the mills cut the wages of cotton-mill operatives by 12.5 percent. Unrest and dissatisfaction were widespread as cotton manufacturers introduced a "10-loom system," stretching out weavers' work. Textile workers were never well organized, and unions at the time represented mostly skilled workers. In response to these work-place changes, however, twenty-six thousand operatives from thirty-three different textile manufacturing companies joined to stage a bitter six-month strike that ended, as most strikes did, in the defeat of the workers. To everyone's credit, there was no violence during the strike and not a single arrest. The strike ended at the instigation of Governor William Lewis Douglas, who investigated workers' claims and conditions and determined that "existing conditions did not justify higher wages." The textile unions accepted the governor's decision, and operatives returned to work under the 12.5 percent reduction. Shortly afterward, the Textile Council, composed of represen-tatives from the major craft unions in Fall River, asked manufacturers for a conference to discuss local industrial conditions, but they refused; there was nothing to talk about. Six months without work helped to fortify class identity among the young workers, so

FIGURE 9.19. "Sandy Beach. Two girls in foreground about 15." (Hine 4177)

that there was only one answer to the question, "Which side are you on?" Industrial conflicts sharpened the lines of division between management and the labor community and shaped the contours of the working child's class consciousness. The 1904 strike was the last citywide strike in Fall River history and left many feeling that there should never be another strike in Fall River if it could be prevented by negotiations.[19]

## THE CHILDREN OF LAWRENCE, MASSACHUSETTS,
## AND THE BREAD AND ROSES STRIKE

New England children working in the textile industry experienced their share of strikes and lockouts, but no labor action was as monumental as the "Bread and Roses" strike in Lawrence in 1912. Reform often helped to ameliorate conflict but did not remove it as a central feature of working-class life. Hine conducted intensive photographic work in Lawrence just prior to the outbreak of the strike, which consumed the lives of everyone in that city and drew the attention of much of the country. Whether the children that Hine photographed were later participants in the upheaval is uncertain (fig. 9.20), but children did play a unique role in the epic conflict. Strikes were one of the arenas in

FIGURE 9.20. "In Washington Mill." (Hine 2460)

which class consciousness was forged, and during the Bread and Roses strike children in Lawrence gained a working knowledge of the principles of class solidarity.

When Hine went to Lawrence in early 1912, it was not to photograph a strike. He was not a photojournalist or a news photographer. He traveled to Lawrence to conduct extensive survey work there, as he had elsewhere, and to document the names and faces of the young workers in the textile center. Hine photographed many groups of child laborers in Lawrence and again demonstrated his remarkable ability to identify and record, however briefly, the stories of his subjects. Take, for instance, a photograph of a crowd of boys (fig. 9.21) in which Hine effectively uses the large group portrait to highlight the youth of some of its members, reminding the viewer in his caption that all the children, even the youngest, work in the local mills. Not only was Hine able to gather the boys together and photograph them—no small feat—but he also identified many of the children in the crowded image, reporting their names and their addresses in his caption. A cursory reading of ethnicities shows a very diverse group of workers. One of the most noted qualities of the Bread and Roses strike was the workers' ability to join in solidarity as one big union under the ideology of "an injury to one is an injury to all." Solidarity enabled the diverse workforce to act as one.

The children were cheap labor. As such, they helped to keep their parents' wages down, just as paying immigrant workers less in wages set them as competitors and suppressed the wages of native-born workers, and paying women less had the tendency to lower male workers' wages. All of this was a reflection of the textile industry's constant drive to reduce its labor costs. Despite the tensions within the working class, Hine's photographs show unity amid the diversity. In images like the one of a group of young workers going into the Ayer Mill early on the morning of September 11, 1911, young and old waited for the morning bell to ring and the day to begin.[20] They all worked under the same labor regime. In working-class communities, child labor was the norm, so when labor conflict struck, children experienced the hardships along with everyone else. In the Bread and Roses strike, labor organizers assigned the children a much more dramatic and central role.

The Lawrence strike committee used a tactic that was little known in the United States but familiar in many European cities—sending the workers' children to stay with strike supporters in other cities for the duration of the strike. The committee believed that "without the cries of hungry children to cause surrender the strikers would win." The children went to the homes of sympathizers who wanted to show solidarity with the strikers while also lifting some of the burden off of strikers' families.

FIGURE 9.21. "All in photos worked (even smallest girl and boys)." (Hine 2462)

Socialists and other radicals not only provided much of the support but also used the campaign to generate publicity and gain a wider audience for the strikers. The Italian Socialist Federation initiated the plan, with assistance from the Socialist Party, and each group took 150 children to New York City. The first contingent of children left for New York on February 10, followed by groups going to Philadelphia, and many children were sent through socialist networks to smaller towns in Massachusetts and to Barre, Vermont.[21] Some 400 children in all were sent to supporters outside Lawrence.

The trip to New York was a baptism in solidarity for the children and an induction into the international working-class movement. According to contemporary accounts, the strikers' children played their part in the drama with enthusiasm and pride. The strike committee appointed Giovanni DiGregorio, Oscar Mazzitelli, Carrie Zaikener, Henry Landwirth, and Margaret Sanger to escort the children to New York. They arrived at Grand Central Station to a cheering crowd. DiGregorio and Mazzitelli described the scene for *Il Proletario*, an Italian American radical newspaper:

> When the train finally entered Grand Central Station, the children began to sing the "International" together; they descended and went in an orderly fashion towards the enormous crowd that was shouting words of welcome from outside the station. That was an unforgettable, emotional moment! There were young and old proletarians of various nationalities that felt the brotherly affection, united by the bond of misery and oppression; they were strangers who at that moment felt they belonged to the same big family that was marching on the path to liberation; they were people of different languages who understood each other very well in the symbol of the waving red flags amongst the crowd. Their hearts were beating in unison; the divisions of ideas, nationalities and language had disappeared. The red flag was the only one present: those that indicate division, hatred and war amongst brothers were not there that evening.[22]

Italian socialist Diabeti Massimo recalled "the tears that dripped from the eyes of the unknown mothers, at the arrival of the unknown children" as they were welcomed into the "cradle of fraternal love."[23] Few forces are as powerful as solidarity in action.

Recognizing the propaganda success of these tactics in reaching a larger public and facilitating wider circles of solidarity, manufacturers in Lawrence wanted them stopped. Colonel E. Le Roy Sweetser, leader of the Massachusetts National Guard, wrote the strike committee on February 17 with a promise that "while I am in command of the troops in Lawrence, I will not permit the shipping off of little children away from their parents to other cities unless I am satisfied that this is done with the consent of the parents of said children." After a few more children were sent to Bridgeport, Connecticut, on February 22, the local marshal declared an end to the exodus. When a group of children tried to leave Lawrence two days later, the police moved in and arrested thirty people, most of them for "congregating," though five women were held for "neglect of children." Some of the children were detained as well, but after hearing just a few of the cases the presiding judge sent them all to a specially appointed committee, with the result that all of the charges were eventually dismissed or were "continued indefinitely."[24]

The unveiled threat to take children away from their parents fortified the parents' determination to fight what they saw as a base violation of their rights. Strikers had taken great care to secure an identification card and parental consent form for each child leaving the city. The police let a sizable group go to Philadelphia after insisting on a complete list of the children who were leaving. One side pushed and the other side pushed back, and in those exchanges, if it didn't already exist, clarity emerged respecting whom the government worked for, as well as whom or what the police and authorities would protect (in this case, property) and whom they would not. Police arrested 296 people in connection with the Lawrence strike, with 220 paying fines and 54 being sentenced to prison; a few cases were dropped, and a few were held over.[25] The strikers charged the police with brutality, stating that the

> crimes of the police during this trouble are almost beyond human imagination. They have dragged young girls from their beds at midnight. They have clubbed the strikers at every opportunity. They have dragged little children from their mothers' arms and with their clubs they have struck women who are in a state of pregnancy. They have placed people under arrest for no reason whatsoever. They have prevented mothers from sending their children out of the city and have laid hold of the children and the mothers violently and threw the children into waiting patrol wagons like so much rubbish. They caused the death of a striker by clubbing the strikers into a state of violence. They have arrested and clubbed young boys and placed under arrest innocent girls for no offense at all."[26]

The children played their part in the strike and paid a price for their engagement.

For many of the children, the strike was an induction into a working-class culture of solidarity and struggle. In the soup kitchens and through songs, slogans, and shouts, parents were teaching their children the basic lessons of working-class life. They were teaching each other as well, living the spirit of their union ideals. Here were the values they lived by: "Nobody back to work until all go together! Don't break ranks; don't scab!"[27] In a public letter to Massachusetts governor Eugene Foss and Lawrence mayor M. F. Scanlon, the strike committee expressed its complete lack of faith in the government, notifying the two politicians that "the striking mill workers of Lawrence, Mass., do not expect you or any of your subordinates to do them justice." The strike committee continued, explaining the workers' position: "You have all done the bidding of the textile manufacturers and we are becoming accustomed to all the ghastly brutalities, the beating of work, the clubbing of children, and all the infamies that your puppets and service tools have heaped on peaceful people in this community." Having suffered all the "atrocities against ourselves and our kin," the committee concluded, "we will remember, we will never forget, and never forgive."[28] The strike transformed participants of all ages.

The strike introduced the children of Lawrence to the potential power of their participation in the industrial work force. Joseph Ettor, speaking to an audience of striking workers at Lawrence's Franco-Belgian Hall, put forth his belief in syndicalism: to win, workers needed simply to "recognize their own solidarity. They have to do nothing, but fold their arms and the world will stop." The workers' bond of solidarity and purpose was the key to their achieving victory as a class, he declared: "With passive resistance, with the workers absolutely refusing to move, laying absolutely silent, they are more powerful than all the weapons and instruments that the other side have for protection and attack." Ettor reminded his audience of this point again and again: "The policeman's club and the militiaman's bayonet cannot weave cloth. It requires textile workers to do that."[29] Hine's photographs of boys and girls in the textile factories of Massachusetts showed that they too were textile workers. The photographs did not depict their class consciousness, however, or reveal any sign of the agency they may have had in their own lives.

# CHAPTER 10

# Trades and Vocational Education

————

L EWIS W. HINE CONCLUDED HIS career with the National Child Labor Committee
by creating two new types of photographs and adding them to his archive: the
physical examination image and the vocational photograph. His photography continued to serve a professional, middle-class audience, but the character of the work
changed significantly. Hine's early child labor photographs are of a muckraking nature;
they were designed to stir up dirt and publicize the existence and life experiences of
child laborers in the United States. Hine used photography as a lever for social reform,
as a tool to motivate political engagement, and as an instrument for the professional
to expose change in policy or statute. His final photographic work while employed by
the NCLC was more positive in its orientation; it identified what was therapeutic and
restorative and, more important, displayed the appropriate course for children making
their way into the workforce. Hine photographed the new roles of the state in structuring solutions to the child labor problem.

In the reformed society that Massachusetts had become by 1916, boys or girls seeking employment had to obtain a work certificate first. Legislation had tightened standards for the primary requirement, proof of the child's age. Previously, the word of the
parents or a local parish priest had been accepted as adequate proof, but now children
had to have documentation of their date of birth, whether a birth certificate or a bap-
tismal record. Reformers made incremental improvements to children's education and
language requirements over the years, holding firm on the demand that every child
complete grade school. The discussion of industrial education, technical training, and
vocational training grew, as survey after survey made it evident that, although tending
a machine in a factory was a deadening experience, children often chose work over

school, seeing no need for an education beyond rudimentary literacy and civic train-ing. Schools were not meeting the needs of large numbers of children entering indus-try. Work requirements for young laborers effectively expanded the role of the state by pushing government into new areas of responsibility. In addition to providing proof that they were old enough to work, children now had to pass a physical examination designed to weed out those who were physically unfit for work, in the opinion of the state. Hine's photographs of these physical examinations were his foray into the new fields of industrial health and occupation safety (fig. 10.1). In these medical photo-graphs, Hine turned his focus away from the labor to the laborer, concerning himself not with the work itself but with the ability or inability of the child to work, now that the state was exercising the role of gatekeeper to the labor force.

Even when illustrating physical deformity, Hine contextualized his photographs with a medical framing rather than a social one. In these images, he moved out of the social milieu and into the medical one. Through his camera Hine was examining children in an institutional setting and diagnosing individual children. He was also documenting a process in which an expanding state bureaucracy demanded medical data on children and professionals, based on their own expertise, efficiently decided whether to certify an application to labor.

Although Hine was not photographing these children at work, he often raised work issues when making his photographic examination. For example, in his caption for one of several photographs he made of Harry Katz of Boston, Hine steps outside the role of photographer to add his professional diagnosis, noting that the boy has no support for his shorter limb, leaving him "obliged to lean against the wall for support." The boy's body, according to Hine, is "not properly balanced," and if he "continued to do standing work without support of crutch, hip would cause boy constant pain." Hine passes the matter over to his professional colleagues, acknowledging the need for the advice of the "examining physician."[1] This is an example of Hine's comfort around other professionals and his characteristic deference to their expertise.

Hine's photographs recorded a fundamental maturation of child labor reform. He was not photographing individual children to condemn the system of child labor. What child labor remained in the state was considered more manageable, opening an opportunity for reformers to inspect individual workers, not workplaces, and to assess what each worker needed in order to enter the workplace safely while also providing efficient labor. In January 1917 Hine photographed Fannie Bowman, a fifteen-year-old bookkeeper in Boston, and identified the particular ailment that threatened her well-being as a worker (fig. 10.2). It is not an image aimed at condemning work but one made with the goal of fixing conditions to allow this individual to work safely. Hine reports that Fannie has a "postural deformity" that makes the "correct position in work most essential." She can work standing or sitting, Hine proffers, as long as

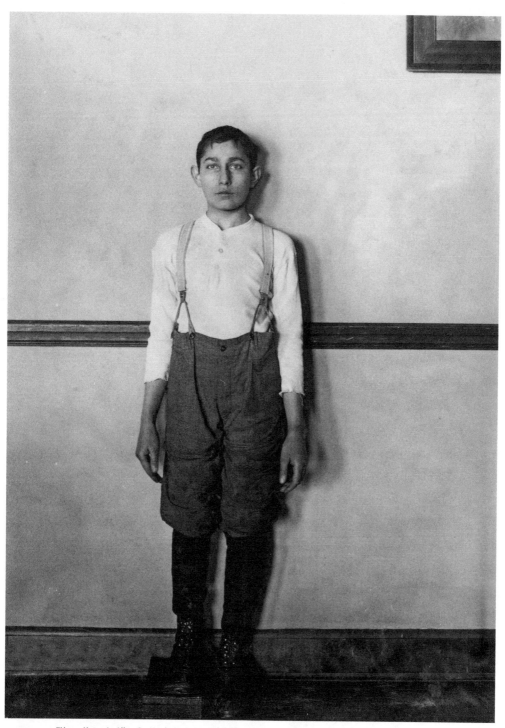

FIGURE 10.1. "Harry Katz, 67 Allen St. Healed T.B. hip." (Hine 4659)

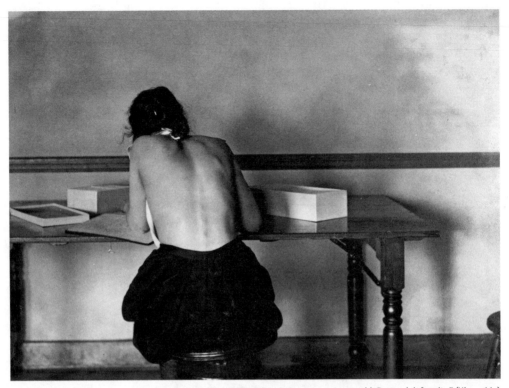

FIGURE 10.2. "Fannie Bowman, 15 years old. Postural deformity." (Hine 4663)

she takes "care to keep good posture." He notes that the "picture shows extremely bad working position for this physical defect"; a bookkeeper with postural deformity should not work in "a hunched, stooped-over position." Hine diagnosed each individual worker with the belief that the worker's particular needs could be tended to.

In his medical photographs, Hine not only identified physical disabilities but also recommended adaptations of the workplace to improve the work environment for that individual, and perhaps for others as well. In addition, Hine noted when health conditions made for a bad fit between an individual and an industry or occupation. This part of Hine's work was about the restorative and rejuvenating impact of the professional class on child labor. Child labor continued to be widespread, but it was brought fully into a regulatory framework controlled by school superintendents, labor bureau inspectors, recreation departments, social welfare organizations, and public health departments. Hine was popularizing the contribution of medical inspection to the regulation of labor. He photographed three girls in an "incorrect sitting position for postural deformity and dorsal curvature cases" (fig. 10.3). It is not healthful, he asserts in the caption, for children with scoliosis to stoop or sit in a "lopsided or humped over

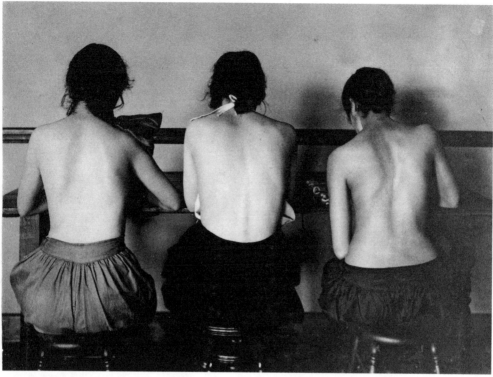

FIGURE 10.3. "Incorrect sitting position for postural deformity and dorsal curvature cases." (Hine 4673)

position." Hine charges that "work in this position is harmful" and calls for the "advice of examining physician." Once again, Hine demonstrates professional deference.

### INDUSTRIAL ACCIDENTS

The photographs of physical examinations are different from the many photographs Hine made of children injured on the job. The physical examination photographs describe health issues and then suggest ways to overcome health limitations and environmental challenges. In the case of his injury photographs, Hine used the images in a more traditional way, to illustrate the idea that children did not belong in the textile industry. At the Union Hospital in Fall River, Hine photographed Estelle Poiriere, a fifteen-year-old doffer in the Granite No. 1 mill who had lacerated the index and middle fingers of her right hand when they were caught in a carding machine (fig. 10.4). When one of the fingers grew stiff, according to Hine, Estelle "had to have cord cut." The injury occurred in December 1915; in mid-June 1916, when Hine made the photograph, Estelle was still an outpatient and had yet to find employment. Hine reported

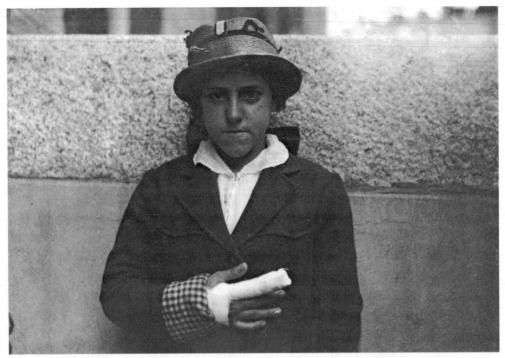

FIGURE 10.4. "Union Hospital case—Estelle Poiriere, 137 Robeson St." (Hine 4302)

Estelle's progress, or lack of progress, from when she was hurt, but what he did not do with this image or with his other traditional images of workplace injuries—in contrast to what he did with the examination photographs—was to recommend changes to the machinery or factory environment. These images were not coupled with suggestions for safety devices on machines, substitute management practices, shorter hours, or any other steps that might be taken to avoid future injuries.

Hine's image of another textile injury follows the same pattern, with Hine high-lighting the injury and its consequences but refraining from suggesting steps to avoid the same sort of injury in the future. Hine photographed seventeen-year-old Molla Mesuretta at the Union Hospital in Fall River (fig. 10.5). Molla had worked in the Sagamore Mill for two years before she was injured while cleaning a loom that was still in motion. Hine was told that the girls "all clean when the machine is going." The image shows a lacerated finger and "partial evulsion of tip of finger and nail." The "finger that she needs most in weaving was hurt," splitting the nail and cutting the fingertip "to the quick." Hine notes the extraordinarily high price of this accident, stating that Molla "will always be handicapped." The finger was still tender, and she was "not able to go back yet." Molla earned $7.50 a week in wages when she was working. Upon

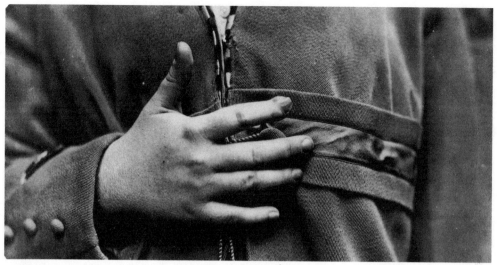

FIGURE 10.5. "Accident case—Union Hospital. Molla Mesuretta." (Hine 4303)

her injury, her employer gave her only $4.67 in compensation. If all the girls continued to clean the machine while it was still moving, additional injuries seemed certain to occur, but there was no incentive for employers to introduce new procedures or safety devices that could potentially reduce the number of these types of factory accidents. As with his other photographs of the victims of workplace accidents, Hine does not make those connections or suggest workplace improvements. He presents Molla as a victim without a clear path to her restoration to the industry.

Workplace injuries were not limited to the young, and on one occasion, Hine photographed an example of the consequences to others of children's carelessness on the job. The photograph shows a large bandaged hand resting in an impromptu sling made from a man's buttoned jacket (fig. 10.6). Hine reports that the man's hand was crushed by a flywheel at the Offset Litho Company when his seventeen-year-old "helper . . . disobeyed instructions and pushed [the] wrong lever." Only the man's torso is visible in the photograph; thus his hand is the only evidence of the accident. This is a different type of image, because it shows (or suggests) a child and an adult who are not in their customary roles. The child, who is not shown, was the victimizer rather than the victim. The injured hand was the result of his carelessness at work.

### INDUSTRIAL EDUCATION

The successes in regulating child labor in Massachusetts resulted in a shift within the reform movement from a focus on showing the dissipation of children at work

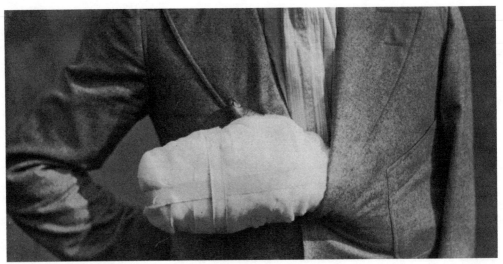

FIGURE 10.6. "Accident case—Man with hand crushed by flywheel." (Hine 4363)

to an interest in creating a healthful, fit, skilled workforce capable of meeting the needs of industry and society. Reformers sought ways to integrate children into the adult workforce without replicating the conditions they were earlier trying to eradicate. Reformers looked to education—industrial, technical, and vocational—for alternatives to children simply quitting school and finding work. One major critique of child labor had been that children entered industry without skills and remained unskilled, giving them little opportunity for advancement. Reformers had put forth many educational variations in the first two decades of the century, but an emphasis on education rather than restriction, on regulation rather than eradication, certainly expressed the new tone of child labor activists. Robert A. Woods, director of the South End Settlement House in Boston, best summed up the sentiment in a speech he made at the Boston Young Men's Christian Union in January 1909: "Industrial education is the positive corollary of the movement against child labor. It is a poor shift to forbid children from entering employment, and provide no influence for making them more effective workers in the meanwhile."[2] Reformers shared in the belief that children would stay in school longer if they were offered courses more relevant to their future as workers.

The overwhelming majority of children in the working-class communities of Massachusetts stopped their education on completing grammar school, at which time they entered the workforce. Woods put the proportion of children in the state who followed that path at between 80 and 95 percent. The calculus was simple, according to Woods: this migration to work was "the result largely of poverty." Poor families needed all the contributions they could gather. Woods, like most who looked at the problem a little more closely, realized that to "a considerable extent it is owing to the unreality of such secondary education as is now offered." The large number of children who could be found in factories rather than classrooms was a product not of ability but of opportunity; as Woods preached, "It has little or no relation to the latent capacities of the rising generation." Children rejected school as soon as the opportunity arose. Their most likely future was to work in the factories their parents toiled in, at the machines their mothers and fathers worked, for the same low wages that justified their joining the workforce in the first place.[3] Woods was by no means alone in this sentiment. Reporting on recent child labor legislation, F. Spencer Baldwin told an audience at the Annual National Child Labor Conference in 1909 that "the truth has been realized that economic incompetency, resulting from lack of definite preparation for self-supporting employment, is in a large measure responsible for pauperism and other social ills."[4] It was necessary to create the right kind of educational opportunities.

The promise of education seemed bright, but the rejection of education did not mean that working-class children were opposed to attending school. They just didn't see the relevance in traditional secondary education. Robert A. Woods argued that parents who often found it impossible to keep their children in school beyond the age of fourteen would make a great effort to enroll them in industrial schools. The benefits of industrial schools, Woods believed, would ultimately "establish for the workman a better standard and give him a longer life." Critics, Woods charged, feared that if too many children were trained to be skilled workers, not enough workers would be left to do unskilled work. Such a danger, Woods contended, was "remote," but if it were to happen, "this would stimulate invention to take the place of unskilled labor. A rise in the price of unskilled labor is a thing to be desired."[5] Raising wages for child laborers would also stimulate invention.

In Massachusetts, where so many children chose to labor rather than to continue with their education after grammar school, state officials, business leaders, social workers, reformers, educators, and others explored alternatives types of education that would better serve working-class children (fig. 10.7). The state legislature authorized a study of the matter, and Governor William L. Douglas put together a Commission on Industrial and Technical Education, appointing Carroll D. Wright, U.S. Commissioner of Labor and former chief of the Massachusetts Bureau of Labor Statistics, to head the commission, with the Honorable Judge Warren A. Reed serving as vice-chair and John Golden, president of the United Textile Workers Union, serving as secretary. The commission then selected Dr. Susan M. Kingsbury, an expert sociologist representing the Boston Women's Educational and Industrial Union, as the principal investigator. Dr. Kingsbury put together a corps of assistants to aid in the investigation. The commission researched the educational needs of children, assessed how those needs were being met by various institutions, and identified new forms of educational effort to be

FIGURE 10.7. "Vocational Printing. Math. class." (Hine 4171)

developed. After a brief but intense period of study, the Douglas Commission submitted its report in May 1905.[6]

Interest in vocational education was widespread in Massachusetts, and Dr. Kingsbury worked hard to solicit the opinions of people in the industrial centers and among the various groups who would be most dramatically affected by changes in the state's educational practices. Dr. Kingsbury held twenty public hearings across the state—nine in Boston and one each in Brockton, Fall River, Fitchburg, Lawrence, Lowell, Lynn, New Bedford, North Adams, Pittsfield, Springfield, and Worcester. Large numbers of people turned out for these hearings, particularly in Springfield, North Adams, and Fitchburg. Dr. Kingsbury and her assistants questioned a broad population of 143 citizens, including farmers, manufacturers, and industrial factory workers, about the traditional educational system in Massachusetts.[7] They found out that most students and parents were not satisfied with the public school curriculum. These citizens felt that public education was far removed from the vocational demands of the state's industrial society. They also expressed concern that students weren't being prepared to make a living when they left school, and that those children entering the workforce lacked the skill to meet the needs and demands of the industries. Business advocates of industrial education believed it was an "indispensable means of increasing the efficiency of the labor force of the state and thus heightening the competitive power of the state's industries."[8] Vocational education seemed the best route to those ends.

Manufacturers and labor unions differed in their perspectives on the establishment of trade and vocational schools to supplement the current public school curriculum. In almost all instances, manufacturers showed widespread support for the establishment of industrial schools to produce the skilled workmen they needed. According to the Douglas Commission's report, manufacturers experienced a near-universal want of "industrial intelligence." In addition to seeking a broadening of the curriculum in schools, manufacturers advocated the expansion of trade schools and vocational education at the public's expense. They were confident that a plan could be created and implemented to cultivate industrial intelligence and increase technical skill. Labor unions, on the other hand, were against any plan for publicly funded trade schools, basing their opposition "on the fear that [the schools] would furnish workmen in numbers sufficiently large to affect the labor market, and bring about a lowering of wages." In the aftermath of their bitter defeat in the 1904 textile strike, unionists were

FIGURE 10.8. "Class in English for employees. Pocasset Mill." (Hine 4174)

also concerned that in times of labor conflict, these schools could be used to furnish workers ready to take the place of union men. The labor unions declared themselves "totally and unalterably opposed" to such schools, which they called "scab hatcheries."[9] The unions' generally cautious perspective could be understood in the context of their fear that industrial education might be turned against labor.

Hine photographed many forms of educational effort designed to supplement grammar school education. One type of educational offering was the evening class, which enabled workers to attend school after their day's labor was done. Because the number of foreign-born among the New England working class was so large, English language classes were common. Hine photographed an English language class for employees of the Pocasset Mill in Fall River (fig. 10.8). The photograph shows a mixed-age group of students sitting at a long table, the students seemingly working at their own individual pace: some are reading, the boy in the forefront is writing, and an older student is speaking with the teacher, who stands over him. This photograph differs from more traditional classroom images in which all the students are roughly the same age. Hine chose the boy laborer closest to the camera lens as his point of focus, suggesting that the boy is valuable to the industry and as such is being treated much like the adult students.

Institutions endeavoring to provide some measure of vocational training already existed at this time. The state joined with cities in establishing textile schools in a few of the major textile centers (fig. 10.9). Textile schools opened in Lowell in 1897, in New Bedford in 1899, and in Fall River in 1904. Each school served the needs of local industry and was supported by annual grants from the state and city, with additional funding coming from student tuition fees. The textile schools taught "all the parts of mill work, the reasons behind the machinery, and all the things that an overseer or a superintendent has to know." Children leaving school for the workplace were informed of these opportunities to learn about the industry when they went for their work certificate. Greater training and education was touted as a path to higher wages and an antidote to the abandoning of children to lives as unskilled workers.[10] Dr. David Snedden, commissioner of education for Massachusetts, stated: "It has been pointed out in the report of the Douglas Commission (of Massachusetts) as well as elsewhere that the period from 14 to 16 is a critical one in the vocational development of large numbers of children. This is the period when economic necessity or ambition tempts children into callings which are temporarily quite remunerative (for children), but which are essentially noneducative. The outcome is the unskilled worker."[11] Young workers who were most in need of education were often enticed more by the wages they received from working.

Most vocational programs were funded by private groups. Some of the skilled trades—masonry, carpentry, plumbing, and sheet metal work, for instance—had ac-

FIGURE 10.9. "Continuation School group at Ipswich Mills." (Hine 4741)

tive apprenticeship programs. The Massachusetts Charitable Mechanic Association conducted an evening trade school for young men, many of whom were already enrolled in apprentice programs. The Women's Educational and Industrial Union of Boston offered day classes in millinery, dressmaking, and sales; some of these classes were highly trade-oriented, while others were more general in their orientation. The Boston Trade School for Girls offered all-day classes in dressmaking, millinery, catering, design, and machine operating, and students' fees and expenses were paid in full through private fundraising (fig. 10.10). The Wells Memorial Institute held free evening classes in electricity, steam and steam engines, mechanical and machine drawing, household science, millinery, dressmaking, and photography; the institute had 1,220 registrants in 1905. Boston's Young Men's Christian Association operated four technical schools—a school of commerce, a polytechnic school, a school of applied electricity and steam engineering, and an automobile school. Each school had numerous departments, and students paid a portion of the instruction costs. The Boston North End Union was a charitable organization that offered plumbing courses in the evening and printing classes during the day. The North Bennet Street Industrial School provided evening classes in sewing and printing to two hundred women and forty men.[12]

FIGURE 10.10. "Continuation school girls looping stocking in Ipswich Mills." (Hine 4738)

Compared with vocational education opportunities in Europe, prospects in Massachusetts for gaining knowledge and skill in a productive industry were "strikingly and painfully inadequate." The Douglas Commission advised Americans to surrender their pride and look seriously at what was going on across the Atlantic, noting that the scope of vocational and technical training in continental Europe "is so broad, its forms are so multifarious, its methods are so scientific, its hold upon public opinions is so complete, the impulse which it is giving to industrial leadership is so powerful, as to entitle it to the most thoughtful and respectful study." European systems for educating child laborers had extensive involvement from both manufacturing and trade associations, much as in the United States. They provided for a much greater number of day and evening schools. What most differentiated vocational education in Europe from that in Massachusetts was the role of government. In Europe, there was a widespread union of national and municipal support and control.[13]

American public schools did not do a good job of preparing children for technical, commercial, mechanical, or agricultural labor. Night schools, an older type of alternative education, still existed in many larger cities but didn't come close to meeting the needs of laboring children. In Germany, the state was much more active in ensuring that working-class children had access to vocational education opportunities. German educators created a new curriculum for children who worked in factories, in stores, or on the farm, an approach that appealed to many American Progressives. Dr. Georg Kerschensteiner, superintendent of schools in Munich, led the way in developing vocational programs that served as a model for Americans. Kerschensteiner stated that the first goal of education for those leaving primary school was the "development of trade efficiency and love of work, and with this the development of those elementary virtues which effectiveness of effort and love of work immediately call forth,—conscientiousness, diligence, perseverance, responsibility, self-restraint, and dedication to a strenuous life." A second aim must also be pursued: "to gain an insight into the relations of individuals to one another and to the State, to understand the laws of health, and to employ the knowledge acquired in the exercise of self-control, justice, and devotion to duty, and in leading a sensible life tempered with a strong feeling of personal responsibility." Many in the United States were attracted to the German system's effort to make good citizens as well as good workers.[14]

Massachusetts followed the lead of German educators in expanding educational opportunities to address the needs of working-class children. In June 1912, the *Child Labor Bulletin* reported that "very recently vocational and continuation schools have been established in various parts of Massachusetts where there is great need for such schools since thirty thousand boys and girls leave the common schools each year at the age of fourteen to take up some trade."[15] In the next few years, child labor activists and educators made significant strides in the provision of vocational and technical training. In 1916, the Massachusetts Child Labor Committee published *A Constructive Program for Securing the Full Benefit of Existing Child Labor Laws in Massachusetts, 1916,* which laid out an extensive plan to make the most of existing laws. The focus of child labor work shifted from placing restrictions on children's admission to wage labor to supporting efforts to uplift those who made the choice to leave school for work. The economy was undergoing changes that required a greater knowledge base and sufficient numbers of skilled workers in new industries such as electricity, chemistry, and auto mechanics. The United States' entry into the war in Europe resulted in an extreme demand for skilled workers. Education became more vital to industries, spurring business leaders to promote both vocational training and the development of a skilled workforce. At the same time, employers also implemented Taylorist practices that deskilled workers and curtailed the need for knowledgeable and highly paid (and often organized) skilled workers. For instance, during World War I, skilled machinists were critical to the war effort and in high demand. When production needs decreased after the war and Taylorist practices turned machinists into machine tenders, more jobs were open to semi- and unskilled workers. In addition to expanding educational opportunities, the more significant long-term result was to deskill industrial operations, allowing for an enormous influx of women into the machine trades.[16]

For Hine, the emphasis on vocational training changed the nature of his work. Hine began taking many more photographs of uplift than of hardship. Instead of trying to depict what was wrong about children's relationship to work, he captured what was right about it. Hine photographed the right fit between the child worker and a trade or a machine, rather than the ill-fitting presence of unskilled labor, fated only for a future of poverty. Hine also had to transform how he portrayed work. Hine photographed the child dissipated by premature labor less and less often between 1915 and when he left the National Child Labor Committee in 1917. He had always recognized the dignity in his subjects, despite the conditions they might have suffered, but now he sought to capture the dignity of both the laborer and the labor. This transformation of his child labor work laid the groundwork for Hine's subsequent corporate photography of the idealized worker. But Hine's vocational images are among the least known of his photographs; only his work in Europe for the American Red Cross at the end of World War I has received less attention.[17]

Hine's vocational photographs were published in many places in 1917, as Hine was completing his service with the National Child Labor Committee. In January 1917 the NCLC published *More Education Pays: A Brief Account of Children in Industry with Descriptions of Certain Common Productive Processes; A Syllabus for the Use of Teachers,* a pamphlet that featured Hine's photographs. That same year the Massachusetts Child Labor Committee published Hine's photos in a sixteen-page booklet, *Out to Win,* by Al Priddy. (Priddy had previously published an autobiography, *Through the Mill,* in which he recounts a boyhood spent working in the cotton mills of New England.) The same intimacy as existed before between Hine and his subjects was still in evidence in these publications; what was new was that Hine was photographing not to condemn but to celebrate. For instance, two photographs that Hine created at Harry C. Taylor's embossing shop on Boston's Court Street show a fifteen-year-old girl (fig. 10.11) and a fourteen-year-old boy learning to work the embossing machine.[18] In Hine's earlier photographs he might have condemned the encounter as dangerous to the children. In these two images, however, the message is much more positive. The child workers are attentive to the lessons being taught by Harry Taylor, who is appropriately watching over his young apprentices, teaching them but also observing as they learn by doing. The gray-haired Taylor appears highly patient and invested in their work. Although neither child is (yet) a master of the craft or of the embossing machine, Hine's vocational photographs illustrate what was then a new ideal for children in the working world; they showed children moving into trade schools and apprenticeships and learning valuable skills to progress within a trade, rather than remaining untrained, unskilled, and unlearned. The images represent a civilizing of industry with regard to children. Children could take part in the industrial world, but only under proper conditions and supervision, as in Harry Taylor's shop.

Hine's vocational images show that Hine was not looking for examples of transgression but rather was seeking to document instances of inclusion in photographing children at work. The motivating idea was that work was no longer a negative force but a positive one. Hine made several of this type of image, in which a young man tends a machine, no longer overwhelmed by it but standing alongside it as a professional. This was true even when the children were not as neatly attired as the boy and girl operating the embossing machine in Taylor's shop. Hine photographed a fifteen-year-old boy "boring at a lathe" in Fall River (fig. 10.12). The boy does not appear to be overwhelmed or frightened by his machine. Hine did not include contextual elements in the image to put his subject matter in a positive light as he often did to cast a subject in a negative light. But what Hine did not display in a photograph he often put in the caption, telling the viewer in this instance that the boy was enrolled in a vocational class. By establishing an educational framework, Hine released the viewer from responsibility. The reformer did not need to stand between the worker and employment. The reformer could take pride that another lad was on the road to future well-being. The vision had

FIGURE 10.11. "Embossing shop of Harry C. Taylor, 61 Court Street." (Hine 4697)

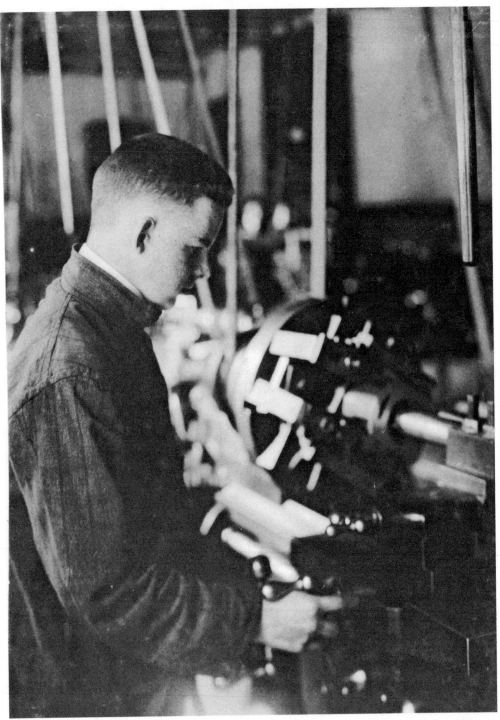

FIGURE 10.12. "[Vocational class. Boring at a lathe. 15 years old.]" (Hine 4164)

changed from protector to promoter of the new educational schemes. Hine's caption helped the viewer to see a demonstration of learning in the image, rather than exploitative child labor.

Some of Hine's images appear dangerous at first glance, until we learn of the educational context, as with Hine's photograph of a boy working at a regular forge. The image shows the boy holding tongs in one hand and a hammer in the other (fig. 10.13). The absence of gloves might signify the resistance of working-class men to wearing protective gear, as workers today might do. More significant is Hine's captioning that the boy is drawing out a lathe tool, indicating that he is making his own tools, perhaps for an apprenticeship as a machinist. Boy laborers such as this one were pursuing admirable pathways as craftsmen and skilled workers. Were they to succeed in becoming machinists, they would be among the best-paid workers in the community. Their status and power was enhanced by the demand for their skills during World War I; members of the International Association of Machinists (IAM) not only won an eight-hour workday but also found themselves courted into the middle class with the lure of home ownership and automobiles.[19]

Educational contexts validated the presence of girls and boys in the workplace.

FIGURE 10.13. "Regular Forge—Drawing out a lathe tool." (Hine 4160)

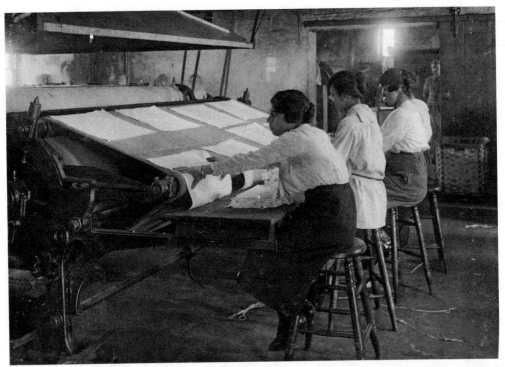

FIGURE 10.14. "Girls working at mangle in Bonanno Laundry, 12 Foster Wharf." (Hine 4731)

They were no longer trespassers. In Hine's vocational photographs, children are safe in the workplace and presumed to be progressing toward the attainment of skills that promised to better their lives. Education and the acquisition of skills sanctioned workplaces that may otherwise have seemed unhealthy. Consider Hine's photograph of girls working at a mangle in the Bonanno Laundry on Foster Wharf in Boston (fig. 10.14). The viewer might not associate working at a mangle with having skills in the same way that a lathe was firmly associated with skills. From the photograph we can't tell whether the four young women were exploited at the laundry. The caption reassures the viewer of the workers' safety, however, and sanctions their work by linking it to education, as Hine notes that the fifteen-year-olds were all attending a continuation school.

Vocational education was one of the only paths to progressing in an occupation or trade because acquiring the skills necessary to move up by experience alone was extremely difficult. Members of the Massachusetts Commission on Industrial Education were highly critical of the methods of manufacturers. The problem, charged the commission, was clear: no one was nurturing the development of young workers for middle or lower supervisory positions. It was difficult at best to obtain the skills and perspective

FIGURE 10.15. "Rhea Quintin—14 years old. Hand drawing in on Webb frame." (Hine 4254)

necessary for a supervisory role. In recommending the establishment of, and increased support for, vocational schools, the commission stated that because of contemporary methods of manufacturing, it was "impossible for the men in the lower positions to fit themselves for higher positions from their shop experience alone."[20] The commission argued that education should provide such men with a route for advancement.

Hine photographed children engaging in that process of self-betterment and acquiring new skills to improve their standing within the workforce. In Fall River he photographed fourteen-year-old Rhea Quintin, who, for the previous three months, had been learning the art of drawing in by hand to set up the warp of the loom (fig. 10.15). In his caption Hine recognizes the "great deal of mental application and accuracy and good oversight" involved in the work, which generally took "over a year to learn." He also reports that Rhea seemed "very young" when she came to get her work certificate, leading Miss Smith in the certificate office to believe that Rhea was simply "a little school girl coming for some other purpose." Hine does not elaborate on what that other purpose might be.

The textile industry workforce was often considered to be unskilled, but there were a small number of skilled positions, one of which was mule spinner. Children moved into those positions through a variety of informal connections—family, friends, one's own kind who knew someone. For children who didn't have the necessary family connections, the textile schools opened the door to the skilled trades, teaching students every aspect of the industry, from working as a sweeper to being a superintendent. The Massachusetts Child Labor Committee promoted the schools' pecuniary value to workers' earnings, promising that "a factory boy or girl who will attend one of these schools has the chance to learn enough to obtain one of the better paying jobs in the factory."[21] The National Child Labor Committee launched a campaign to persuade children that "It pays to stay in school."

Hine photographed children preparing for a trade by more traditional means that were unconnected to the new vocational opportunities, such as learning a family business. Twelve-year-old Frank De Natale worked in his father's barbershop on Hanover Street in Boston. Hine photographed Frank wielding a straight razor as he shaved one of his father's customers (fig. 10.16). Frank worked in the shop after school and on Saturdays. Hine reports nothing more about Frank's situation, but judging from the confidence that the customer and Frank's father seem to have placed in the boy, we might assume that the young De Natale had at least developed a steady hand by that point in his barbering career.

In another image of children joining the family business, Hine shows the Messina brothers working together. In the photograph, fifteen-year-old Vincenzo and his eleven-year-old brother, Angelo, bake bread for their father's Cambridge bakery (fig. 10.17). Vincenzo is putting dough in the oven with a bread paddle while Angelo stands

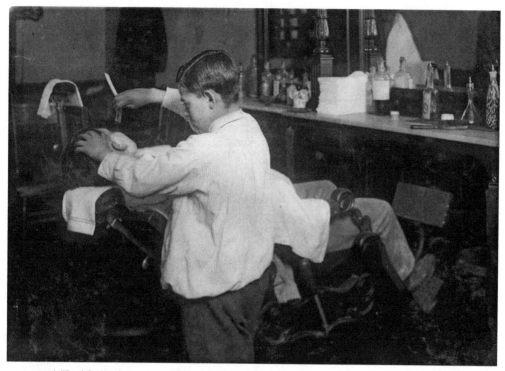

FIGURE 10.16. "Frank De Natale, a 12-year old barber." (Hine 4729)

at his side holding dough balls; working as a helper to his brother, Angelo is learning the trade. Angelo undoubtedly broadened his contributions as he grew up, with Vincenzo and other family members or employees mentoring him. Hine captions that Vincenzo already performs a man's labor, working a twelve-hour overnight shift. Angelo contributes a great deal as well, Hine notes, by tending the store and helping with the baking. In the case of this photograph as well as that of Frank De Natale, the fact that it was the father's business contextualizes the image and the trade being practiced—barbering and baking, both common to and welcome in every neighborhood.

Hine photographed other children who were entering the workforce and taking on occupations outside of the usual textile or shoe shop but who were not necessarily being prepared for any particular skill. Waitressing, for example, was an occupation that was growing rapidly, but while children could certainly learn the nuances of the job, no matter how much they learned, opportunities for advancement were not there. Hine took a photograph of fourteen-year-old Delia Kane as she waited tables at the Exchange Luncheon in South Boston (fig. 10.18). Delia is wearing a clean white outfit and carrying a tray of food. She momentarily rests the tray on a counter upon which sits a large coffee urn, serving both functionally as a dispenser of coffee and in the

FIGURE 10.17. "Vincenzo Messina, 15 years old and brother Angelo, 11 years old, baking bread for father." (Hine 4710)

photograph as a symbol of Delia's occupation. In certain respects, this image was very much like one of Hine's traditional work portraits: a photograph of a worker with tool in hand. In this case, the coffee urn and the food tray replace the bobbin as a symbol of labor. Hine acknowledged the opening up of occupations in the economy, but he regretted that some young women exchanged one limited position for another. Delia may have thought differently. Did the white shirt and skirt provide added social or cultural compensation? Were the hours better? Did the Exchange Luncheon pay more?

Not surprisingly, young women faced greater limitations than men did in both their educational and employment opportunities. Hine conducted photographic research for inclusion in *Working Girls in Evening Schools: A Statistical Study*, a report by Mary Van Kleeck, secretary of the Committee on Women's Work of the Russell Sage Foundation. The images were of workers in New York City, but the message was the same there as in New England: industrial training presented women with unique challenges. As Van Kleeck stated in her report, "It is generally conceded that if the problem of industrial training be a difficult one for boys, it is still more so for girls."[22] One problem, Van Kleeck suggested, was that the public expected boys to work for wages but did not attach the same expectation to girls, although modern life was making it

FIGURE 10.18. "Exchange Luncheon. Delia Kane, 14 years old." (Hine 4712)

increasingly necessary for girls to earn a wage. Another difficulty women encountered was that female occupations were changing rapidly as technology and cultural conventions evolved. The ultimate pressure facing female educational opportunities was women's constant struggle to move beyond traditional home pursuits such as sewing or cooking. Van Kleeck quoted the president of the National Women's Trade Union League as stating that "no denial of trade education will keep a girl out of a trade, and if she is denied entrance by the front door as a skilled, trained artisan, she will enter by the back door as an under-bidder."[23] Here once again was an expression of trepidation that trade education might be used to pit one part of the workforce against another.

Vocational educators like David Snedden, the first commissioner of education for Massachusetts, and his colleague Charles Prosser, whom Snedden appointed as the state's deputy commissioner for vocational education, advocated reforming the public educational system to include greater vocational training. Snedden promoted social efficiency through vocationalism. Influenced by the practices of scientific management, particularly Taylorism, Snedden maintained that education in an industrial society should be better aligned with industrial tendencies and practices in order to create greater efficiencies. Snedden and Prosser argued that schools existed to prepare children for the economic world that awaited them. Given that the two men wanted practical education designed for realistic occupational opportunities, they looked to business for leadership in creating a more efficient educational system that would meet the needs of the great mass of students, who were headed to work at the soonest possible moment. Prosser formed an advocacy organization, the National Society for the Promotion of Industrial Education (NSPIE), to lobby the federal government for support for vocational education. The NSPIE and Hoke Smith, a former governor of Georgia and a Democratic senator, together wrote several important pieces of legislation and crafted a political coalition to secure passage of the 1917 Smith-Hughes Act, a major victory for the vocationalists. The Smith-Hughes Act made the federal government a partner in the provision of vocational education.[24]

Many Progressive educators rejected the style of vocational training promoted by Snedden and Prosser and debated with other reformers the proper goals of education. Should schools be a conduit to industry, or should education be a force for self-realization and personal development? Should educators design schools to fit the needs of industry, or should children be taught to become active, civically engaged citizens who can then make their way into industry and shape the direction of industry, as leading Progressive educator John Dewey advocated? The growing collusion between educators and the business world led Benjamin Gruenberg, a New York teacher writing in the *American Teacher* (later to become the official journal of the American Federation of Teachers), to argue against business models of efficiency:

> The organization and the methods of the schools have taken on the form of those commercial enterprises that distinguish our economic life. We have yielded to the arrogance of "big business men" and have accepted their criteria of efficiency at their own valuation, without question. We have consented to measure the results of educational efforts in terms of price and product—the terms that prevail in the factory and the department store. But education, since it deals . . . with individualities, is not analogous to a standardizable manufacturing process. Education must measure its efficiency not in terms . . . of so many student-hours per dollar of salary; it must measure its efficiency in terms of increased humanism, increased power to do, increased capacity to appreciate.[25]

The business model of education sought standardized results rather than acknowledging and encouraging the variations in human ability.

John Dewey believed that education should nurture intelligence and that intelligence would enable the individual's successful adaptation to the conditions of life. Dewey's perspective was essentially an inversion of Snedden's or Prosser's. Education was not about managing society in the interests of industry but about creating intelligent individuals who could direct the economy according to the needs of society. Dewey saw the Smith-Hughes Act as narrowly conceived legislation that would passively promote class polarization by restricting a student's career or life goal and constraining the host of educational influences a student might encounter that could broaden his or her understanding of life.[26]

Where Hine fell along this educational divide is not entirely clear. Hine was a student of Dewey's while enrolled at the University of Chicago, but he was also a photographer who made many of the images used in the publications of the vocational education movement. Hine's photographs of children in educational contexts are, like the vast majority of his photographs, respectful of their subjects and humanistic in their ability to illustrate the human condition with optimism tempered by basic honesty. The Massachusetts Child Labor Committee hired Hine to make photographs that would promote the state's vocational programs. Hine fulfilled his assignment, producing dozens of photographs of industrial educational settings, but in doing so, he also left behind his role as a social photographer seeking reform. The vocational work called forth a very different type of subject: a child civilized by a reformed industrial society. Look at Hine's photograph of a boy working with sheet metal in a prep technology class that met for six periods a week at the technical high school in Fall River (fig. 10.19). The young man is well dressed in his work jacket, starched collar, and tie; his face is clean; he appears to be well groomed and engaged in something purposeful. This is not an image of something *to be* reformed but an image of something that *is* reformed. This photograph depicts the domestication of the working class and its integration into society; it is a photograph of industrial education, well ordered and disciplined, rationalized with business logic, and valued not because it educated for democracy or citizenship but because it educated for corporate capitalism.

The evolution of Hine's photography mirrors the trajectory of Progressive reform. When Hine began photographing child laborers for the National Child Labor Committee in 1908, his mission was simple: photograph to reveal, uncover, muckrake, and discover through the "scientific" social survey the conditions of children laboring across this country. His findings were visual tools to help people understand what was happening in modern society, to make sense of the phenomenal changes wrought in the early decades of the century. He saw his work as a lever for social change. Joined by a vast network of like-minded professionals, Hine played the unintended role of illustrating the creation of the Progressive state. His photographs condemned society, but they also recorded changes in the direction of the child labor reform movement and shifts in the currents of Progressive reform at the local, national, and international level. The individuals and organizations that Hine connected with as he traveled the country were gestating places of reform. Hine came into communities with an agenda, but his success was dependent on local individuals engaged in the issue of child labor. They showed him where to look and sometimes what to look for in his search for evidence of indignities to children. Hine was at the nexus of the exchange of information between

FIGURE 10.19. "Sheet metal class." (Hine 4165)

local and national constituencies concerned about child welfare. His photographs also reflect the efforts of business and the new middle-class professional to rationalize and humanize capitalism. Business leaders could be genuinely committed to reform for humanistic reasons, while others were committed to achieving economic and social efficiency. Middle-class professionals, of whom Hine was one, worked very hard in behalf of working-class children. They strove to improve the conditions of child workers and to enhance, while also containing, working-class children's life opportunities. They did so, however, within their own class, gender, racial, and cultural limitations, so there was no shortage of cultural insensitivity, as demonstrated by well-meaning but patronizing attacks on the reformed and their families and, in many cases, on their way of life. If or when order did not prevail, the state was practiced in the use of force against labor; for instance, the Massachusetts National Guard was sent to Lawrence in 1912 in an attempt to crush the "Bread and Roses" strike. During World War I, labor upheaval and labor discipline both increased. Military conscription was one way of disciplining labor. What happened in Lawrence in 1912—the militarization of the city during the strike—became the more general condition of labor during and after the war, as patriotism took over from reason. In 1917 Hine went overseas to work for the Red Cross.

Lewis W. Hine left the National Child Labor Committee for many reasons, not the least of which was that the NCLC reduced his hours and his pay. In a larger sense, his departure from the NCLC was a symptom of his success. Political and economic winds were shifting, and as the United States came closer to entering the war in Europe, the drive to secure sufficient skilled labor to meet the crisis remained a prominent and persistent issue. If this was a war to make the world safe for democracy, there were many at home who would have appreciated a share of the victory booty. Hine entered the child labor movement when generosity of head and heart prevailed; he left it as that spirit gave way to the push for wartime efficiency. Throughout his work for the NCLC, Hine recognized and even celebrated the pluralist traditions in our history that during World War I gave way to the more intolerant forces of one hundred percent Americanism. Hine began his career with the NCLC by investigating and displaying the dissipation of child workers. He ended his work in child labor where he would take up after the war, photographing workers and the industrial world they had mastered. The corporate portraits that Hine created in the postwar period had their roots in this final period of Hine's child labor work. Hine's photographs evolved from documenting the child at work to evaluating the child's ability to work. Hine photographed the child measured and weighed, and just before he left the National Child Labor Committee, he learned to photograph the child learning a trade. The immensity of Lewis W. Hine's contribution to our knowledge of the lives of working-class children is a gift of great magnitude. It is a legacy worthy of his subjects.

# Notes

1. The truth was a little more complex. Almost 48 percent of Vermonters lived in areas that the 1910 U.S. census designated as urban, while fewer than 53 percent lived in rural areas. Historical Census Browser, Geospatial and Statistical Data Center, Census of 1910; see http://mapserver.lib.virginia.edu.

2. In May 1910 Hine made a few photographs of the Henry Street Settlement in New York. His last extensive survey work was in December, photographing children working in New Jersey glass factories. See the National Child Labor Committee collection, Library of Congress.

3. Lewis W. Hine photograph and caption numbers 1049, 1050, and 1052.

4. No. R-219. Joint Resolution Commemorating the U.S. Postal Service's Issuance of the Addie Laird Child Labor Postage Stamp, Acts of the 1997–1998 Vermont Legislature; see http://leg.state.vt.us/docs.

5. Elizabeth Winthrop, "Through the Mill," *Smithsonian Magazine* (September 2006); Elizabeth Winthrop, *Counting on Grace* (New York: Wendy Lamb Books, 2006); U.S. Bureau of the Census, *Thirteenth Census of the United States, 1910,* Pownal Town, County Bennington, VT, sheet 12B, May 4, 1910. See also the Vermont Women's History Project at http://womenshistory.vermont.gov.

6. George Dimock, "Priceless Children: Child Labor and the Pictorialist Ideal," in *Priceless Children: American Photographs 1890–1925,* ed. George Dimock (Greensboro, NC: Weatherspoon Art Museum, 2001), 15; Sarah E. Chinn, *Inventing Modern Adolescence: The Children of Immigrants in Turn-of-the-Century America* (New Brunswick, NJ: Rutgers University Press, 2009), 63. To make Addie's portrait less significant because she did not commission it is to misunderstand the character of this working-class archive that offered only temporal visibility upon the stage of history.

7. Alan Trachtenberg, *Reading American Photographs: Images as History, Mathew Brady to Walker Evans* (New York: Hill and Wang, 1989), 170–72.

8. Lewis W. Hine, "Photography in the Schools," *Photographic Times* 40, no. 8 (August 1908); John Berger, *Ways of Seeing* (London: Penguin Press, 1977).

9. Walter Rosenblum, foreword to *America and Lewis Hine: Photographs 1904–1940* (1977; repr. New York: Aperture, 1997), 11.

10. Daile Kaplan, ed., *Photo Story: Selected Letters and Photographs of Lewis W. Hine* (Washington, DC: Smithsonian Institution, 1992), 60–61.

11. Robert H. Wiebe, *The Search for Order, 1877–1920* (New York: Hill and Wang, 1966).

12. Lewis W. Hine, "Social Photography; How the Camera May Help in the Social Uplift," *Proceedings of the National Conference of Charities and Correction at the Thirty-Sixth Annual Session held in the City of Buffalo, New York, June 9–16, 1909*, ed. Alexander Johnson (Fort Wayne, IN: Press of Fort Wayne, 1909), 355–59; Lewis W. Hine, "Advertisement," *Charities and the Commons*, June 6, 1908.

13. Maren Stange, *Symbols of Ideal Life: Social Documentary Photography in America 1890–1950* (Cambridge: Cambridge University Press, 1989), 47–85.

14. Hine, "Social Photography," 355.

15. Ibid., 356. Susan Sontag, in her powerful work *On Photography*, argued that "moralists who love photographs always hope that words will save the picture." Captions, Sontag cautioned, "do tend to override the evidence of our eyes; but no caption can permanently restrict or secure a picture's meaning." Sontag continued with her exploration, noting that "what the moralists are demanding from a photograph is that it do what no photograph can ever do—*speak*." The caption, she argued, was the photograph's "missing voice, and it is expected to speak for truth. But even an entirely accurate caption is only one interpretation, necessarily a limiting one, of the photograph to which it is attached." Susan Sontag, *On Photography* (New York: Picador USA, 2001), 107–9.

16. Hine, "Social Photography," 356–57.

17. Steven Mintz, *Huck's Raft: A History of American Childhood* (Cambridge: Harvard University Press, 2004), 2–3; Christopher Lasch, *Haven in a Heartless World: The Family Besieged* (New York: Basic Books, 1977).

18. Edward T. Devine, "The New View of the Child," National Child Labor Committee pamphlet no. 71 (1908), 1–7.

19. James Guimond, *American Photography and the American Dream* (Chapel Hill: University of North Carolina, 1991), 55–98.

20. Miles Orvell, *American Photography* (New York: Oxford University Press, 2003), 14.

21. Laura Wexler, *Tender Violence: Domestic Visions in an Age of U.S. Imperialism* (Chapel Hill: University of North Carolina Press, 2000), 208.

22. Wexler, *Tender Violence*, 59.

23. Kate Sampsell-Willmann, *Lewis Hine as Social Critic* (Jackson: University Press of Mississippi, 2009), 8–10.

24. Edward T. Devine, *The Spirit of Social Work* (New York: Russell Sage, 1911), v–vi.

25. Roland Barthes, *The Rustle of Language* (Oxford: Blackwell, 1986), 141–48.

26. David Montgomery casually mentions child labor in his magnum opus, *The Fall of the House of Labor*, but even he doesn't make the obvious connection between child labor and the dilution of skilled trades. Montgomery, *The Fall of the House of Labor: The Workplace, the State, and American Labor Activism, 1865–1925* (New York: Cambridge University Press, 1987). See also Cecelia F. Bucki, "Dilution and Craft Tradition: Munitions Workers in Bridgeport, Connecticut, 1915–19," in *The New England Working Class and the New Labor History*, ed. Herbert G. Gutman and Donald H. Bell (Urbana: University of Illinois Press, 1987).

27. Karl Marx, *Capital: A Critique of Political Economy*, ed. Frederick Engels (New York: Modern Library, 1906), 431.

28. *Report of the Secretary of the Treasury, [Alexander Hamilton,] on the Subject of Manufactures, Made the 5th of December, 1791.*

29. U.S. Bureau of the Census, *Twelfth Census of the United States, 1900* (Washington, DC: U.S. Government Printing Office, 1900).

30. Sampsell-Willmann, *Lewis Hine as Social Critic*, 20–21.

31. G. Stanley Hall, *Adolescence: Its Psychology and Its Relations to Physiology, Anthropology, Sociology, Sex, Crime, Religion and Education* (New York: D. Appleton, 1904); Dorothy Ross, *G. Stanley Hall: The Psychologist as Prophet* (Chicago: University of Chicago Press, 1972). See also Chinn, *Inventing Modern Adolescence*, 32–33.

32. George Dimock, "Children of the Mills: Re-Reading Lewis Hine's Child-Labour Photographs," *Oxford Art Journal* 16, no. 2 (1993): 37–54; Stange, *Symbols of Ideal Life*; Trachtenberg, *Reading American Photographs*.

33. Rosenblum, *America and Lewis Hine*, 12.

34. Alan Trachtenberg, "Lewis Hine: The World of His Art," in *Photography in Print: Writings from 1816 to the Present*, ed. Vicki Goldberg (Albuquerque: University of New Mexico Press, 1981), 250.

35. Dimock, "Children of the Mills," 37–54; Dimock, "Priceless Children"; Vicki Goldberg, *Lewis W. Hine: Children at Work* (New York: Prestel, 1999); Judith Mara Gutman, *Lewis W. Hine and the American Social Conscience* (New York: Walker, 1967); Elizabeth McCausland, "Boswell of Ellis Island: What Statisticians Forgot to Record about the Crowded Gateway to a New World Is Geographically Remembered in Photography of Lewis W. Hine," *U.S. Camera*, January–February 1939, 58–62; Jennifer L. Peresie, "Crusader with a Camera: Lewis Hine and His Battle against Child 'Slavery,'" *Pennsylvania History* 23 (Summer 1997): 4–13; *America and Lewis Hine*; Russell Freedman, *Kids at Work: Lewis Hine and the Crusade against Child Labor* (New York: Clarion Books, 1994); Sampsell-Willmann, *Lewis Hine as Social Critic*; Peter Seixas, "Lewis Hine: From 'Social' to 'Interpretive' Photographer," *American Quarterly* 39, no. 3 (1987): 381–409; Stange, *Symbols of Ideal Life*; Trachtenberg, *Reading American Photographs*. One of the exceptions is Chinn, *Inventing Modern Adolescence*. See also John R. Kemp, ed., *Lewis Hine: Photographs of Child Labor in the New South* (Jackson:

University Press of Mississippi, 1986); Joseph D. Thomas, "Lewis Hine: Portrait of Two Cities—Fall River and New Bedford," *Spinner* 3 (1984): 6–27.

36. I move this shift from social to interpretive photographer earlier, to Hine's work on vocational trades in Boston. See Seixas, "Lewis Hine," 381–409; Trachtenberg, *Reading American Photographs*, 221.

37. Henry Barnard, *Legal Provision Respecting the Education and Employment of Children in Factories, &c.; With Examples of Improvement in Manufacturing Districts* (Hartford, CT: Case, Tiffany and Burnham Printers, 1842).

38. *Brooklyn Eagle,* September 15, 1902, 4; *Brooklyn Eagle,* November 19, 1902, 5. See also May Wood Simmons, "Education in the South," *American Journal of Sociology* 10, no. 3 (November 1904): 382–407.

39. U.S. Bureau of the Census, *Twelfth Census* and *Thirteenth Census; Robert Macieski, "Cities and Suburbs in New England," in *The Encyclopedia of New England,* ed. Burt Feintuch and David Watters (New Haven: Yale University Press, 2005).

40. Guimond, *American Photography,* 55–98; Edgar Gardner Murphy, *Problems of the Present South: A Discussion of Certain of the Educational, Industrial, and Political Issues in the Southern States* (New York: Macmillan, 1905), 309–29.

41. Florence Kelley, "New England's Lost Leadership," *Annals of the American Academy of Political and Social Science* 35 (March 1910): 150–51.

42. Barbara Dayer Gallati, *Children of the Gilded Era: Portraits by Sargent, Renoir, Cassatt, and Their Contemporaries* (New York: Merrell, 2004); Ronald G. Pisano, *William Merritt Chase: The Complete Catalogue of Known and Documented Work by William Merritt Chase,* vol. 2, *Portraits in Oil* (New Haven: Yale University Press, 2007), 43; Anne Higonnet, *Pictures of Innocence: The History and Crisis of Ideal Childhood* (London: Thames and Hudson, 1998); *New York Times,* April 9, 1901. Background on Käsebier's *Blessed Art Thou among Women* can be found at the Metropolitan Museum of Art website, www.metmuseum.org.

43. Stephen Nissenbaum and Dona Brown, "Changing New England, 1865–1945," in *Picturing Old New England: Image and Memory,* ed. William H. Truettner and Roger B. Stein (New Haven: Yale University Press, 1999), 4.

44. Alan Axelrod, *Colonial Revival in America* (New York: Norton, 1985).

45. See Robert Hariman and John Louis Lucaites, *No Caption Needed: Iconic Photographs, Public Culture, and Liberal Democracy* (Chicago: University of Chicago Press, 2007), for an interesting analysis of iconic images in popular culture.

46. Allan Sekula, "The Body and the Archive," in *The Contest of Meaning,* ed. Richard Bolton (Cambridge: MIT Press, 1989), 358.

47. Ibid., 345, 373.

## 2. STREET TRADES

1. Helene Silverberg, "'A Government of Men': Gender, the City, and the New Science

of Politics," in *Gender and American Social Science: The Formative Years,* ed. Helene Silverberg (Princeton, NJ: Princeton University Press, 1998), 156–57.

2. Samuel P. Hays, *Conservation and the Gospel of Efficiency: The Progressive Conservation Movement, 1890–1920* (Pittsburgh: University of Pittsburgh Press, 1999); M. Christine Boyer, *Dreaming the Rational City: The Myth of American City Planning* (Cambridge: MIT Press, 1986); Paul Boyer, *Urban Masses and Moral Order in America, 1820–1920* (Cambridge: Harvard University Press, 1978); Daphne Spain, *Gendered Spaces* (Chapel Hill: University of North Carolina Press, 1992); Ann Douglas, *Purity and Danger: An Analysis of the Concepts of Pollution and Taboo* (London: Routledge, 1984). The historian Peter C. Baldwin argues that reformers believed that the "dark wisdom" of the streets, knowledge of the world gained in urban public spaces, was "unsuited to young people." Their concerns about children in the streets, he believed, "revealed cultural conflicts between middle-class and working-class Americans over the meanings of modern night, and over the raising of children." Baldwin, "'Nocturnal Habits and Dark Wisdom': The American Response to Children in the Streets at Night, 1880–1930," *Journal of Social History* 35, no. 3 (2002): 593.

3. Charles Mulford Robinson, one of the principal advocates of the City Beautiful movement in America, cited the belief of some city planners that streets could have "social influence," while recognizing that the influence of the streets could cut either way: "If environments can enliven and inspire, they must also have the power to deaden and discourage." In short: "Mean streets make mean people." Robinson, "The Sociology of a Street Layout," *Annals of the American Academy of Political and Social Science* 51 (Housing and Town Planning Conference, January 1914): 196; Udetta D. Brown, "A Brief Survey of Housing Conditions in Bridgeport, Connecticut: Investigation and Report for the Bridgeport Housing Association" (March–May 1914); John Nolen, "More Houses for Bridgeport: Report to the Chamber of Commerce, Bridgeport, Connecticut" (August 1916); John Nolen, "Better City Planning for Bridgeport: Some Fundamental Proposals to the City Plan Commission (1916)"; Carol Aronovici, "Housing Conditions in New Haven" (Civic Federation of New Haven, January 1913); Henry P. Fairchild, "An Industrial Survey of a New Haven District" (Civic Federation of New Haven, April 1913); "The Report and Recommendations of the Bridgeport Vice Commission" (1916); "Report of the Hartford Vice Commission, Hartford, Conn." (1913).

4. National Child Labor Committee, Owen Lovejoy, General Secretary, *Fifth Annual Report* (New York, September 1909).

5. One popular example is the Ragged Dick series. Horatio Alger Jr., *Ragged Dick: Or, Street Life in New York with the Boot-Blacks* (Philadelphia: Henry T. Coates, 1895).

6. Lillian D. Wald, *The House on Henry Street* (New York: Holt, 1915), 147–48.

7. Edward N. Clopper, *Child Labor in City Streets* (New York: Macmillan, 1912), 17.

8. Clopper, *Child Labor in City Streets,* 19–20.

9. Philip Davis, *Street-Land: Its Little People and Big Problems* (Boston: S. J. Parkhill, 1915), xvi–xviii. Davis used Hine's photographs to illustrate his book.

10. Laura Wexler, *Tender Violence: Domestic Visions in an Age of U.S. Imperialism* (Chapel Hill: University of North Carolina Press, 2000), 50.

11. *Thirteenth Census of the United States, 1910—Population*, Hartford, CT, enumeration district 157 (May 3, 1910).

12. Education was always an important theme in Hine's photographs. A few years after his Connecticut work, when Hine was put in charge of National Child Labor Committee exhibitions, he created a poster titled *At the University of Experience* that summarized the numerous threats posed by learning in the street. This university had a distinguished faculty that included President Mephistopheles in the "Department of Immoral Philosophy"; Professor Fagin in the "Department of Purse Snatching and Burglary"; Professor Mercury, a "Specialist in Deception and Theft"; and Professor Bacchus, an "Intoxication Expert." In the center of the poster were three of Hine's photographs of street life, labeled "Views of Campus." In the Elementary Course of Study, students learned about truancy, profane language, lying, deception, unwholesome food, irregular hours, bad weather, defiance of parental control, stimulants, cigarettes, and liquors. The advanced program of study covered theft, vicious associates, evil resorts, the ethics of prison life, disease through contact with vice, murders, scandals, and "all sensations of the day." The poster concluded with a restatement of the republican ideal, reminding the public that "You Expect Him to Be President," but it countered that with a more realistic assessment of the likely future for graduates of this university: "The Most Delinquent Boys Come from Street Trades." Hine photograph 3741.

13. Also see Hine photograph 0603, of Tommy De Lucco, nine years old, a newsie who had been selling papers for two years, sometimes working until 8:00 in the evening.

14. Kathy Peiss, *Cheap Amusements: Working Women and Leisure in Turn-of-the-Century New York* (Philadelphia: Temple University Press, 1986). The modern city could cruelly "over-stimulate" the senses of girls like Mary, according to Jane Addams, who argued that, in the street, "the newly awakened senses are appealed to by all that is gaudy and sensual, by the flippant street music, the highly colored theater posters, the trashy love stories, the feathered hats, the cheap heroics of the revolvers displayed in the pawn-shop windows." This "fundamental susceptibility" of the young, Addams asserted, was "thus evoked without a corresponding stir of the higher imagination, and the result is as dangerous as possible." Addams, *The Spirit of Youth and the City Streets* (New York: Macmillan, 1909), 27.

15. Hine photograph 0655.

16. Scott Nearing, "The Newsboy at Night in Philadelphia," *Charities and the Commons* 17 (February 2, 1906), quoted in Clopper, *Child Labor in City Streets*, 135.

17. *Thirteenth Census of the United States, 1910*, Hartford, April 22, enumeration district 290, sheet 10.

18. Mabel A. Wiley, "A Study of the Problem of Girl Delinquency in New Haven" (Civic Federation of New Haven, March 1915).

19. Clopper, *Child Labor in City Streets*, 117.

20. John Spargo, *The Bitter Cry of the Children* (New York: Macmillan, 1906), 185.

21. Spargo, *The Bitter Cry of the Children*, 185. Messengers in Connecticut were indeed highly familiar with the local prostitutes, criminals, and characters of the underworld. They went so far as to recommend prostitutes, available in different types of brothels and at various prices, to H. M. Bremer of the National Child Labor Committee. They also demonstrated their familiarity with the diverse requests for food, drink, drugs, or anything else the women might need. H. M. Bremer, "Investigation of the Messenger Services of Connecticut," National Child Labor Committee Collection papers, box 4 (May 1914), Library of Congress.

22. Clopper, *Child Labor in City Streets*, 102–3.

23. Florence Kelley, *Some Ethical Gains through Legislation* (1905; repr., New York: Arno, 1969), 18.

24. Wald, *The House on Henry Street*, 148–49.

25. Kelley, *Some Ethical Gains*, 13; Clopper, *Child Labor in City Streets*, 12–13.

26. Clopper, *Child Labor in City Streets*, 80–81.

27. School Document No. 15, 1909, Boston Public Schools, 36, quoted in Clopper, *Child Labor in City Streets*, 212–13.

28. Kelley, *Some Ethical Gains through Legislation*, 14–15.

29. Hine photograph 4645.

30. Walter Rosenblum, foreword to *America and Lewis Hine: Photographs 1904–1940* (New York: Aperture, 1977; repr., 1997), 13.

31. Lewis W. Hine, "Vermont Street Trades, etc.," December 1916, National Child Labor Committee Collection papers, box 4, Library of Congress.

32. Mary Ritter Beard, *Woman's Work in Municipalities* (New York: D. Appleton, 1915), 133.

33. Beard, *Woman's Work in Municipalities*, 133.

34. Everett G. Hill, *A Modern History of New Haven and Eastern New Haven County*, vol. 1 (New York: S. J. Clarke, 1918), 41–43.

35. Davis, *Street-Land*, xvi.

36. Hine photograph 0629.

37. Davis, *Street-Land*, xvi–xviii.

## 3. TEXTILES

1. Barbara M. Tucker, *Samuel Slater and the Origins of the American Textile Industry, 1790–1860* (New York: Cornell University Press, 1984).

2. In *Confronting Decline: The Political Economy of Deindustrialization in Twentieth-Century New England* (Gainesville: University Press of Florida, 2013), David Koistinen states that the decline of traditional industries in New England began in the 1920s, but in fact it began in the nineteenth century with capital flight from North to South; see Beth English, *A Common Thread: Labor, Politics, and Capital Mobility in the Textile Industry* (Athens: University of Georgia Press, 2006), 1–5. In his book *Working-Class Community*

*in Industrial America: Work, Leisure, and Struggle in Two Industrial Cities, 1880–1930* (Westport, CT: Greenwood Press, 1979), John T. Cumbler notes that the industrial workforce in New England began to decline during the first decade of the twentieth century, despite continued growth in cotton textiles in Fall River, Massachusetts. Cumbler attributes the workers' decline to the introduction of the Northrup power loom, which was part of a necessary effort to extract greater production in order to compete with mills in the South. Cumbler argues that child labor was central to survival in Fall River and became even more critical as conditions began to deteriorate.

3. A. J. McKelway, "The Mill or the Farm?," in National Child Labor Committee, *Child Employing Industries: Proceedings of the Sixth Annual Conference, Boston, Massachusetts, January 13–16, 1910* (New York, 1910), 55.

4. Dr. Stephen S. Wise, "Justice to the Child," in National Child Labor Committee, *Child Employing Industries,* 36–37.

5. J. Howard Nichols, letter to the editor, *Boston Evening Transcript,* October 30, 1901.

6. Samuel McCune Lindsay, "Unequal Laws an Impediment to Child Labor Legislation," in National Child Labor Committee, *Child Employing Industries,* 16–17.

7. *Child-Labor Bill: Hearings on H.R. 12292, A Bill to Prevent Interstate Commerce in the Products of Child Labor, and for Other Purposes, Before the Committee on Labor,* 63rd Congress, Second Session (May 22, 1914) (statement of Lewis W. Parker), in *The Child and the State: Select Documents,* ed. Grace Abbott, vol. 1, *Legal Status in the Family; Apprenticeship and Child Labor* (Chicago: University of Chicago Press, 1938), 85–90. See also Shelley Sallee, *The Whiteness of Child Labor Reform in the New South* (Athens: University of Georgia Press, 2004), for an excellent treatment of race and child labor reform in the South.

8. Shelton Stromquist, *Reinventing "The People": The Progressive Movement, the Class Problem, and the Origins of Moderns Liberalism* (University of Illinois Press 2006), 95.

9. Hon. Curtis Guild Jr., "Child Labor Legislation in Massachusetts," in *Child Employing Industries,* 9.

10. Hine photograph 0674.

11. Lewis W. Hine, "Child Labor Conditions in New England," National Child Labor Committee Collection, Library of Congress, 1909, 1.

12. Ibid.

13. Ibid., 2.

14. Ibid., 2–3.

15. U.S. Bureau of Labor, *Report on Condition of Woman and Child Wage-Earners in the United States* (Washington, DC: Government Printing Office, 1910), 1:402; English, *A Common Thread,* 74.

16. Hine, "Child Labor Conditions in New England," 3.

17. Hine photograph 0762.

18. Ibid.

19. Ibid., 3–4.

20. Ibid., 4; Hine photograph 0753.

21. Hine, "Child Labor Conditions in New England," 4–5.

22. Women's Educational and Industrial Union, "A Group of Working Children in Manchester, New Hampshire" (1916), 4–5.

23. In April 1915, one of the students in the Vocational Advising Department of the Women's Educational and Industrial Union, located in Boston, conducted an investigation into the home and industrial conditions of a select group of working children in Manchester, New Hampshire. She selected 115 children for this investigation; these children constituted 16 percent of the boys and girls who left school and went to work at fourteen or fifteen years of age between February 1, 1913, and March 1, 1915. In each case, she obtained information regarding the child's school training. The superintendent supplied names and addresses of children and additional information "under the school record" for the study. The State Department of Public Instruction sent "facts" about the number of children working in the local corporations. Investigators interviewed teachers, employers, parents, and the working children themselves for "data concerning the family and family income, the attitude of children and parents toward school, and the children's past and present occupations, and desires." The children in the study closely resembled the children Hine photographed. Ibid., 2.

24. Molly Ladd-Taylor, *Mother-Work: Women, Child Welfare, and the State, 1890–1930* (Urbana: University of Illinois Press, 1994).

25. Women's Educational and Industrial Union, "A Group of Working Children," 5.

26. Ibid., 6.

27. Ibid., 10–13.

28. Amoskeag Manufacturing Company Employee Records, Fred Normandin, October 14, 1911, and October 21, 1911, at the Manchester Historic Association Archives.

29. Hine, "Child Labor Conditions in New England," 5.

30. H.R. Rep. No. 2745 (1906).

31. A. Maurice Low, "On Woman and Child Labor," *Boston Globe*, August 10, 1909.

32. Daile Kaplan, ed., *Photo Story: Selected Letters and Photographs of Lewis W. Hine* (Washington, DC: Smithsonian Institution Press, 1992), 7.

## 4. EXHIBITING CHILD LABOR

1. See Hine's work in the six volumes of the Pittsburgh Survey published from 1909 to 1914: (vol. 1) Elizabeth Beardsley Butler, *Women and the Trades, Pittsburgh, 1907–1908* (1909; repr., Pittsburgh: University of Pittsburgh Press, 1984); (vol. 2) Crystal Eastman, *Work-Accidents and the Law* (1910; repr., New York: Arno, 1969); (vol. 3) John A. Fitch, *The Steel Workers* (1910; repr., Pittsburgh: University of Pittsburgh Press, 1989); (vol. 4) Margaret F. Byington, *Homestead: The Households of a Mill Town* (1910; repr., New York: Arno, 1969); (vol. 5) *The Pittsburgh District Civic Frontage*, ed. Paul U. Kellogg (1914; repr., New

York: Arno, 1974); (vol. 6) *Wage-Earning Pittsburgh,* ed. Paul U. Kellogg (1914; repr., New York: Arno, 1974).

2. Elspeth H. Brown, "Labor, Management and Photography as 'Social Hieroglyphic': The National Cash Register Company and the Social Museum Collection," in *Instituting Reform: The Social Museum of Harvard University, 1903–1931,* ed. Deborah Martin Kao and Michelle Lamunière (Cambridge: Harvard Art Museums, 2012; distributed by Yale University Press), 203. Brown writes about how "capitalist visuality" worked as moderator between workers and their employers at the National Cash Register Company and formed the basis for corporate welfare.

3. Julie K. Brown, "Making 'Social Facts' Visible in the Early Progressive Era: The Harvard Social Museum and Its Counterparts," in Kao and Lamunière, *Instituting Reform,* 93–109.

4. Julie K. Brown, *Contesting Images: Photography and the World's Columbian Exposition* (Tucson: University of Arizona Press, 1994), xiii, xvi.

5. See Shawn Michelle Smith, *American Archives: Gender, Race, and Class in Visual Culture* (Princeton, NJ: Princeton University Press, 1999) and *Photography on the Color Line: W. E. B. Du Bois, Race, and Visual Culture* (Durham, NC: Duke University Press, 2004), for rich analysis of DuBois's "American Negro" exhibit for the 1900 Paris Exposition. See also Deborah Willis's *Reflections in Black: A History of Black Photographers, 1840 to the Present* (New York: Norton, 2000).

6. Lois H. Silverman, *The Social Work of Museums* (New York: Routledge, 2010), 11; Jacob A. Riis, *The Battle with the Slum* (Mineola, NY: Dover Publications, 1998), 140–43.

7. Allen H. Eaton and Shelby M. Harrison, *A Bibliography of Social Surveys* (New York: Russell Sage Foundation, 1930), xvi; Jacob A. Riis, *How the Other Half Lives: Studies among the Tenements of New York* (New York: Charles Scribner's Sons, 1890); Residents of Hull House, *Hull-House Maps and Papers: A Presentation of Nationalities and Wages in a Congested District of Chicago* (New York: Thomas Y. Crowell, 1895); W. E. B. Du Bois, *The Philadelphia Negro: A Social Study* (Philadelphia: University of Pennsylvania, 1899). See also the published findings of the Pittsburgh Survey, cited in note 1.

8. Eaton and Harrison, *Bibliography of Social Surveys,* xxiv.

9. *"1915" Boston Exposition Official Catalogue and the Boston-1915 Year Book* ([Boston]: "1915" Boston Exposition Company, 1909), iv.

10. "Boston—1915," *Civic League Bulletin of Newport, R.I.* 4, no. 5 (January 1910): 6.

11. "Boston: 1915," *National Magazine* 30, no. 6 (September 1909): 696–98.

12. In the first room of the massive exhibit, for instance, was a copy of the central part of the display created by George Carroll Curtis for the 1900 Paris World's Fair, a topographical model of metropolitan Boston. Following its Paris showing, the original exhibit was then housed in the Agassiz Museum at Harvard University. *"1915" Boston Exposition Official Catalogue,* 20.

13. *New England Magazine: An Illustrated Monthly* 43 (September 1910–February 1911): 667.

14. John L. Sewall, "Boston-1915 and Its Child Welfare Work," in *Proceedings of the Child*

*Conference for Research and Welfare 2, Held at Clark University, Worcester, Mass., June 28–July 2, 1910* (New York: G. E. Stechert, 1910), 22.

15. Ibid., 22–28.

16. *New Boston Pageant Program: A Monthly Record of Progress in Developing a Greater and Finer City* 1, no. 7 (November 1910), 15–16.

17. *"1915" Boston Exposition Official Catalogue*, xii, 14.

18. *New Boston Pageant Program*, 16–17. See also David Glassberg, *American Historical Pageantry: The Uses of Tradition in the Early Twentieth Century* (Chapel Hill: University of North Carolina Press, 1990).

19. See the work of Wallace Nutting throughout New England and his reverence for the Puritan century.

20. *"1915" Boston Exposition Official Catalogue*, 54–55.

21. Hine's photographs from his October 1909 trip to Boston are among his best as far as capturing on film the problems and the promise of city life. Hine photographed a wide array of street work, scavenging activity, and ordered institutional imagery to serve the purpose of the "1915" Boston exhibit. He was driven by the question "Where should we be in 1915?" in selecting images for the exhibit, with the result that a majority of his images in the exhibit portrayed opportunities for uplift and improvement. Many of the photographs that he left off the exhibit list were equally as powerful as the ones he displayed.

22. Hine photograph 0949.

23. See Judith E. Smith, *Family Connections: A History of Italian and Jewish Immigrant Lives in Providence, Rhode Island, 1900–1940* (Albany: State University of New York Press, 1985), and Ardis Cameron, *Radicals of the Worst Sort: Laboring Women in Lawrence, Massachusetts, 1860–1912* (Urbana: University of Illinois Press, 1993).

24. See Hine photographs 0887, 0891, 0896, and 0915 in particular for more charged imagery.

25. See Hine photograph 0948 for an image of a corps of truant officers in Boston.

26. In addition to the photographs Hine created specifically for the "1915" Boston Exposition, photographs from his work on the Pittsburgh Survey were also on display. Room 48 had an exhibit showing the results of investigations carried out in Pittsburgh under the auspices of the Russell Sage Foundation. This exhibit was organized into eleven sections: (1) "The outline of the survey"; (2) "Physical—Administrative—Social"; (3) "Make-up of population"; (4) "New dwellings"; (5) "Types of workers"; (6) "Mortality"; (7) "Typhoid"; (8) "Women in the stogie industry"; (9) "Industrial accidents"; (10) "Housing"; and (11) "The Pittsburgh district." *"1915" Boston Exposition Official Catalogue*, 60.

27. Chicago School of Civics and Philanthropy, *City Welfare: Aids and Opportunities*, Bulletin no. 13 (October 1911): 9–11.

28. Ibid., 11; Francis Greenwood Peabody, *The Social Museum as an Instrument of University Teaching* (Cambridge: Harvard University, 1908), 3–4.

29. Peabody, *The Social Museum*, 4–7.

## 5. SARDINES

1. Jane E. Radcliffe, "Perspectives on Children in Maine's Canning Industry, 1907–1911," *Maine Historical Society Quarterly* 24, no. 4 (1985): 362–63, 366–67.

2. "Labor Laws of Maine," *Twenty-First Annual Report of the Bureau of Industrial and Labor Statistics for the State of Maine* (Augusta, ME: Kennebec Journal Print, 1907), 472.

3. Radcliffe, "Perspectives on Children," 367–68.

4. Ernest Stagg Whitin, "Children in the Canning Industry," *Outlook* (January 21, 1905): 177–79.

5. John Spargo, *The Bitter Cry of the Children* (New York: Macmillan, 1906), 170–71; Radcliffe, "Perspectives on Children," 373–74.

6. Radcliffe, "Perspectives on Children," 374–75; Hugh D. Hindman, *Child Labor: An American History* (New York: M. E. Sharpe, 2002), 249.

7. "Women and Children in Sardine Factories," *Twenty-First Annual Report of the Bureau of Industrial and Labor Statistics for the State of Maine,* 121; Radcliffe, "Perspectives on Children," 375.

8. "Women and Children in Sardine Factories," 123.

9. Ibid., 130, 137.

10. Ibid., 128.

11. Ibid., 129.

12. George Brown Goode, *The Fisheries and Fishery Industries of the United States* (Washington, DC: U.S. Government Printing Office, 1887), section 5, vol. 1, 510; quoted in Radcliffe, "Perspectives on Children," 377–378. See also pages 78–89 in the *Fourteenth Annual Report of the Bureau of Industrial and Labor Statistics for the State of Maine* (Augusta, ME: Kennebec Journal Print, 1900) for a description of the various stages of sardine processing.

13. Ernest Stagg Whitin, *Factory Legislation in Maine* (PhD diss., Columbia University, 1908), 133–36.

14. Lewis W. Hine, "Photographic Investigation of Child Labor Conditions in Sardine Canneries of Maine, August, 1911," for the National Child Labor Committee, 1911, Manuscript Division, Library of Congress, 1.

15. Ibid.

16. Ibid., 1–2.

17. Ibid., 4.

18. Ibid., 4.

19. Hine photograph 2459.

20. Hine photograph 2443.

21. Hine, "Photographic Investigation of Child Labor Conditions in Sardine Canneries," 5.

22. Ibid., 5.

23. Ibid., 6.

24. Ibid., 3.

25. Ibid., 4–5.

26. Ibid., 6.

27. Ibid., 5–7.

28. Ibid., 7.

## 6. FARM AND SEASONAL LABOR

1. Ruth McIntire, "Children in Agriculture," National Child Labor Committee pamphlet no. 284 (New York: National Child Labor Committee, February 1918). Agriculture continued to operate outside the protection of child labor legislation even after passage of the New Deal's Fair Labor Standards Act in 1938. See Hugh D. Hindman, *Child Labor: An American History* (New York: M. E. Sharpe, 2002), 286–90.

2. Owen R. Lovejoy, "Progress in Child Labor Reform: Seventh Annual Report for the Fiscal Year Ending September 30, 1911" (New York: National Child Labor Committee, 1911), 7.

3. Hine photographs 3962, 3964.

4. William A. McKeever, *Farm Boys and Girls* (New York: Macmillan, 1913), 2–6.

5. Hine photograph 3993.

6. Hine photograph 3989.

7. See the discussion of Spargo's *The Bitter Cry of the Children* in chapter 5.

8. Lovejoy, "Progress in Child Labor Reform," 7.

9. Henry S. Griffith, *History of the Town of Carver, Massachusetts: Historical Review 1637 to 1910* (New Bedford, MA: E. Anthony and Sons, 1913), 217–20.

10. U.S. Immigration Commission, "Immigrants in Industries (in Twenty-Five Parts): Part 24: Recent Immigrants in Agriculture," vol. II (Washington, DC: Government Printing Office, 1911), 539–42.

11. See Hine photographs 2549, 2550.

12. Lewis W. Hine, Richard K. Conant, and Owen R. Lovejoy, "Child Labor on the Cranberry Bogs of Massachusetts," National Child Labor Committee Collection, Library of Congress Manuscripts (1911), 2–3, 5.

13. U.S. Immigration Commission, "Immigrants in Industries," 539.

14. Marilyn Halter, *Between Race and Ethnicity: Cape Verdean American Immigrants, 1860–1965* (Urbana: University of Illinois Press, 1993), 92, 99.

15. Matthew Frye Jacobson, *Barbarian Virtues: The United States Encounters Foreign Peoples at Home and Abroad, 1876–1917* (New York: Hill and Wang, 2000).

16. Halter, *Between Race and Ethnicity*.

17. Hine, Conant, and Lovejoy, "Child Labor on the Cranberry Bogs of Massachusetts," 2–3, 5.

18. See Ida B. Wells-Barnett, *The Red Record: Tabulated Statistics and Alleged Causes of Lynching in the United States* (1895; repr., Gloucester, UK: Dodo Press, 2009), for the gruesome details.

19. U.S. Immigration Commission, "Immigrants in Industries," 547, 550, 551. In 1904 a Brava who had picked cranberries on a Sunday was found guilty of violating the Sabbath after

the presiding judge refused to allow the jury to consider the question of whether the defendant's gathering of cranberries had been "a work of necessity." The Supreme Judicial Court of Massachusetts later affirmed the lower court's decision; see "Commonwealth vs. Edwin M. White," in *Massachusetts Reports (190): Cases Argued and Determined in the Supreme Judicial Court of Massachusetts, December 1905–March 1906, Henry Walton Swift, Reporter* (Boston: Little, Brown, 1906), 578–82.

20. U.S. Immigration Commission, "Immigrants in Industries," 551.

21. See the film *Ethnic Notions,* directed by Marlon Riggs (1986; San Francisco: California Newsreel, 1987).

22. Hine, Conant, and Lovejoy, "Child Labor on the Cranberry Bogs of Massachusetts," 1.

23. Ibid., 3–4.

24. Ibid., 3, 5.

25. Ibid., 3–4.

26. See Hine photograph 2583.

27. Hine, Conant, and Lovejoy, "Child Labor on the Cranberry Bogs of Massachusetts," 3.

28. Ibid., 5.

29. Hine photograph 2564.

30. Hine, Conant, and Lovejoy, "Child Labor on the Cranberry Bogs of Massachusetts," 3–4.

31. Ibid., 2–3, 5.

32. See also Hine photograph 2556.

33. Josiah C. Folsom, "Farm Labor in Massachusetts, 1921," U.S. Department of Agriculture bulletin no. 1220 (Washington, DC, April 1924).

34. Hine, Conant, and Lovejoy, "Child Labor on the Cranberry Bogs of Massachusetts," 5.

35. Lewis W. Hine, "Children and Tobacco in Connecticut," Report to the National Child Labor Committee (August 1917), National Child Labor Committee papers, Library of Congress, Manuscript Division, 1.

36. Hine, "Children and Tobacco in Connecticut," 1–2.

37. Ibid., 2, 4.

38. Ibid., 1. See also Robert Macieski, "'The Home of the Working Man Is the Balance Wheel of Democracy': Housing Reform in Wartime Bridgeport," *Journal of Urban History* (July 2000); and Cecelia Bucki, "Dilution and Craft Tradition: Munitions Workers in Bridgeport, Connecticut, 1915–19," in *The New England Working Class and the New Labor History,* ed. Herbert G. Gutman and Donald H. Bell (Urbana: University of Illinois Press, 1987).

39. Hine, "Children and Tobacco in Connecticut," 2.

40. Ibid.

41. Harriet A. Byrne, "Child Labor in Representative Tobacco-Growing Areas," U.S. Department of Labor, Children's Bureau, publication no. 155 (Washington, DC: Government Printing Office, 1926), 38, 41.

42. See Hine photograph 4877.

43. Hine photograph 4909.

44. Hine photograph 4869.

45. Hine, "Children and Tobacco in Connecticut," 2–3.

46. Ibid., 3.

47. Ibid.

48. Ibid.

49. Ibid., 3–4.

50. Ibid., 4.

51. Ibid., 6.

52. Hindman, *Child Labor,* 286.

## 7. EXHIBITING CHILD WELFARE

1. *Hand Book of the Chicago Industrial Exhibit* (Chicago: Chicago Industrial Exhibit, 1907), Working Women Collection, Harvard University, 6; Anna Louise Strong, *Child-Welfare Exhibits: Types and Preparation,* U.S. Department of Labor, Children's Bureau, miscellaneous series no. 4, publication no. 14 (Washington, DC: U.S. Government Printing Office, 1915), 7.

2. Currie D. Breckinridge, R.N., "'The Child in the Midst': The New York–Chicago Child Welfare Exhibit," *American Journal of Nursing* 11, no. 10 (July 1911): 815.

3. Anna Louise Strong, "Child Welfare Exhibits," in *The Encyclopedia of Sunday Schools and Religious Education,* ed. John T. McFarland, Benjamin S. Winchester, R. Douglas Fraser, and the Rev. J. Williams Butcher (New York: Thomas Nelson & Sons, 1915), 219.

4. Ibid., 220.

5. Anna Louise Strong, "The Child Welfare Exhibit as a Means of Child Helping," *Proceedings of the National Conference of Charities and Correction, at the Fortieth Annual Session Held in Seattle, Washington, July 5–12, 1913* (Fort Wayne, IN: Fort Wayne Printing, 1913), 311.

6. Strong, "Child Welfare Exhibits," 220–21.

7. Strong, "The Child Welfare Exhibit as a Means of Child Helping," 313–14.

8. Ibid., 314, 312.

9. Ibid., 315.

10. Ibid., 315–16.

11. Strong, "The Child Welfare Exhibit as a Means of Child Helping," 317; Strong, "Child Welfare Exhibits," 221; *Hand Book of the Rhode Island Child Welfare Conference and Exhibit* ([Providence, RI?]: 1913), 2.

12. "Child Welfare Exhibit Planned," *Providence Evening News,* September 24, 1912.

13. U.S. Census Bureau, *Thirteenth Census of the United States, 1910,* Providence Ward 5, April 27, 1910, Enumeration District 97, Sheet 12A; U.S. Census Bureau, *Fourteenth Census of the United States, 1920,* Providence Ward 5, January 8, 1920, Enumeration District 228, Sheet 9B. See also Hine photograph 3163.

14. Hine photographs 3153 and 3154.

15. *Hand Book of the Rhode Island Child Welfare Conference and Exhibit,* 11.

16. Ibid., 13.

17. See Hine photographs 3179 and 3180.

18. Ibid., 7; Evart Grant Routzahn and Mary Swain Routzahn, *The A B C of Exhibit Planning* (New York: Russell Sage Foundation, 1918), 75.

19. *Hand Book of the Rhode Island Child Welfare Conference and Exhibit,* 7.

20. Judith E. Smith, *Family Connections: A History of Italian and Jewish Immigrant Lives in Providence, Rhode Island, 1900–1940* (Albany: State University of New York Press, 1985).

21. U.S. Census Bureau, *Thirteenth Census of the United States, 1910,* Providence Ward 9, Enumeration District 248, Sheet 9.

22. Hine photograph 3215.

23. Routzahn and Routzahn, *The A B C of Exhibit Planning,* 4.

24. Ibid., 71–72.

25. Ibid.

26. *Charlotte Daily Observer,* January 9, 1911, 4.

27. Edwin H. McCloskey, "Free Shows Make Hit," in *Moving Picture World* 25, no. 4 (1915): 693

28. "Pictures Not Justified," *Boston Globe,* July 27, 1915, 14.

29. "Sees No Reason to Change," *Boston Globe,* July 29, 1915, 11.

30. "Cushing Defends Slides," *Boston Globe,* July 28, 1915, 9.

### 8. HOMEWORK

1. Lewis W. Hine, "Tasks in the Tenements," *Child Labor Bulletin* 3, no. 1 (May 1914): 95.

2. Massachusetts Bureau of Statistics, *Industrial Home Work in Massachusetts,* Labor Bulletin no. 101 (Boston: Wright and Potter Printing, 1914), 13.

3. U.S. Department of Labor, Women's Bureau, *Home Work in Bridgeport, Connecticut,* Bulletin no. 9, December 1919 (Washington, DC: Government Printing Office, 1920), 5.

4. Susan Porter Benson, "Women, Work, and the Family Economy: Industrial Homework in Rhode Island in 1934," in *Homework: Historical and Contemporary Perspectives on Paid Labor at Home,* ed. Eileen Boris and Cynthia R. Daniels (Urbana: University of Illinois Press, 1989), 53.

5. Eileen Boris, *Home to Work: Motherhood and the Politics of Industrial Homework in the United States* (Cambridge: Cambridge University Press, 1994), 81–122.

6. Massachusetts Child Labor Committee, *Child Labor in Massachusetts Tenements: Annual Report of the Massachusetts Child Labor Committee, January 1, 1913* (Boston: Massachusetts State Child Labor Committee, 1913), 5–6.

7. Hine, "Tasks in the Tenements," 95.

8. Karl Marx and Friedrich Engels had recognized this decades earlier in *The Communist Manifesto.* Capitalism is a powerful historical force, they claimed, that consumes everything in its path. As capitalism developed, all social relations became part of a cash nexus, and the bourgeoisie tore "away from the family its sentimental veil, and . . . reduced the family relation to a mere money relation." According to Marx and Engels, the need for constantly expanding markets for its products "chases the bourgeoisie over the entire

surface of the globe. It must nestle everywhere, settle everywhere, establish connections everywhere." Karl Marx and Friedrich Engels, *Manifesto of the Communist Party*, trans. Samuel Moore (New York: International Publishers, 1948), 11–12.

9. Florence Kelley, "The Invasion of Family Life by Industry," *Annals of the American Academy of Political and Social Science* 34, no. 1 (July 1909): 90–96. See also *Muller v. Oregon*, 208 U.S. 412 (1908), a landmark Supreme Court decision that reinforced gender stereotypes and justified the use of labor laws in the interests of society. The decision upheld an Oregon State restriction on women's working hours that was predicated on the state's need to protect women's health.

10. Hine, "Tasks in the Tenements," 95–96.

11. Mary Van Kleeck, "Child Labor in Home Industries," in "Child Employing Industries," supplement, *Annals of the American Academy of Political and Social Science* 35 (March 1910): 145–49.

12. Rheta Childe Dorr, "The Child Who Toils at Home," *Hampton Magazine*, April 1912, 183.

13. Ibid., 183–84.

14. Linda Gordon, "The Progressive-Era Transformation of Child Protection, 1900–1920," in *Childhood in America*, ed. Paula S. Fass and Mary Ann Mason (New York: New York University Press, 2000), 548.

15. Massachusetts Bureau of Statistics, *Industrial Home Work in Massachusetts* (1914), 22–23.

16. Ibid., 24.

17. Dorr, "The Child Who Toils at Home," 184.

18. Van Kleeck, "Child Labor in Home Industries," 146.

19. Ibid., 147.

20. Massachusetts Bureau of Statistics, *Industrial Home Work in Massachusetts* (1914), 3.

21. U.S. Department of Labor, Women's Bureau, *Home Work in Bridgeport, Connecticut*; Harriet A. Byrne and Bertha Blair, *Industrial Home Work in Rhode Island, with Special Reference to the Lace Industry*, Women's Bureau Bulletin no. 131 (Washington, DC: Government Printing Office, 1935).

22. Massachusetts Bureau of Statistics, *Industrial Home Work in Massachusetts* (1914), 10–11.

23. U.S. Department of Labor, Women's Bureau, *Home Work in Bridgeport, Connecticut*, 3; Byrne and Blair, *Industrial Home Work in Rhode Island*, v.

24. See Judith Sealander, *The Failed Century of the Child: Governing America's Young in the Twentieth Century* (Cambridge: Cambridge University Press, 2003), for an excellent exploration of the ways in which the state used child welfare as a wedge to enter into the lives of the working class, immigrants, and the poor. See also Kathryn Kish Sklar, "The Historical Foundations of Women's Power in the Creation of the American Welfare State, 1830–1930," in *Mothers of a New World: Maternalist Politics and the Origins of Welfare States*, ed. Seth Koven and Sonya Michel (New York: Routledge Press, 1993), 43–93; and Robyn Muncy, *Creating a Female Dominion in American Reform, 1890–1935* (New York: Oxford University Press, 1991).

25. U.S. Department of Labor, Women's Bureau, *Home Work in Bridgeport, Connecticut*, 5.

26. Massachusetts Bureau of Statistics, *Industrial Home Work in Massachusetts* (Boston: Women's Educational and Industrial Union, 1915), xiv, xx, 12–13.

27. Ibid., 140–42.

28. Massachusetts Bureau of Statistics, *Industrial Home Work in Massachusetts* (1914), 3–4, 20–21.

29. Ibid., 32.

30. Byrne and Blair, *Industrial Home Work in Rhode Island*, 1.

31. U.S. Department of Labor, Women's Bureau, *Home Work in Bridgeport, Connecticut*, 12.

32. See also Hine photograph 2964-A.

33. Massachusetts Bureau of Statistics, *Industrial Home Work in Massachusetts* (1914), 4.

34. Gordon, "The Progressive-Era Transformation of Child Protection," 548.

35. John Tagg, *The Burden of Representation: Essays on Photographies and Histories* (Minneapolis: University of Minnesota Press, 1993), 9.

36. Dorr, "The Child Who Toils at Home," 185–86.

37. Lewis W. Hine photograph 3032.

38. Massachusetts Bureau of Statistics, *Industrial Home Work in Massachusetts* (1914), 41.

39. Dorr, "The Child Who Toils at Home," 186.

40. Massachusetts Bureau of Statistics, *Industrial Home Work in Massachusetts* (1914), 5, 19–21.

41. Dorr, "The Child Who Toils at Home," 188.

42. Tamara K. Hareven's *Family Time and Industrial Time: The Relationship between the Family and Work in a New England Industrial Community* (Cambridge: Cambridge University Press, 1982) does a magnificent job of linking women's work choices to different stages in the family life cycle.

43. Molly Ladd-Taylor, *Mother-Work: Women, Child Welfare, and the State, 1890–1930* (Urbana: University of Illinois Press, 1994).

44. Dorr, "The Child Who Toils at Home," 185.

45. See also Hine photograph 2963-A.

46. "Most of the home work on wearing apparel is distributed directly to the workers. Usually they or their children call at the office, store, or factory from which the work is given out, but in some cases, where materials are exceptionally bulky, the factory sends a team to deliver and collect work at regular intervals." Massachusetts Bureau of Statistics, *Industrial Home Work in Massachusetts* (1914), 63.

47. Hine photograph 2955-A.

48. Hine photograph 2962-A.

49. U.S. Department of Labor, Women's Bureau, "Home Work in Bridgeport," 12; see also Hine photographs 2983-A and 3139.

50. Massachusetts Bureau of Statistics, *Industrial Home Work in Massachusetts* (1914), 34.

51. Byrne and Blair, *Industrial Home Work in Rhode Island*, 10.

52. U.S. Department of Labor, Women's Bureau, *Home Work in Bridgeport, Connecticut,* 9.

53. See also Hine photographs 2949-A and 2950-A.

54. Dorr, "The Child Who Toils at Home," 187.

55. Hine photograph 3030; see also Massachusetts Bureau of Statistics, *Industrial Home Work in Massachusetts* (1914), 5.

56. Massachusetts Bureau of Statistics, *Industrial Home Work in Massachusetts* (1914), 60.

57. Byrne and Blair, *Industrial Home Work in Rhode Island,* 16.

58. Hine photograph 2909.

59. Hine photographs 2951-A, 2952-A, 2953-A, and 2977-B.

60. Massachusetts Bureau of Statistics, *Industrial Home Work in Massachusetts* (1914), 28, 45.

61. Ibid., 6.

62. National Child Labor Committee, *Child Workers in the Tenements,* pamphlet no. 181 (New York: National Child Labor Committee, 1912), 1.

63. U.S. Department of Labor, Women's Bureau, *Home Work in Bridgeport, Connecticut,* 6.

64. Ibid., 7.

65. Massachusetts Bureau of Statistics, *Industrial Home Work in Massachusetts* (1914).

66. Hine, "Tasks in the Tenements," 97.

## 9. WORKING-CLASS COMMUNITIES

1. Some of the works that delve deeply into the history of the New England textile cities, and which I have drawn on in writing this chapter, are Mary H. Blewett, *The Last Generation: Work and Life in the Textile Mills of Lowell, Massachusetts, 1910–1960* (Amherst: University of Massachusetts Press, 1990); Mary H. Blewett, *Constant Turmoil: The Politics of Industrial Life in Nineteenth-Century New England* (Amherst: University of Massachusetts Press, 2000); Stuart Blumin, *The Emergence of the Middle Class: Social Experience in the American City, 1760–1900* (Cambridge: Cambridge University Press, 1989); Ardis Cameron, *Radicals of the Worst Sort: Laboring Women in Lawrence, Massachusetts, 1860–1912* (Urbana: University of Illinois Press, 1995); Christopher Clark, *The Roots of Rural Capitalism: Western Massachusetts, 1780–1860* (New York: Cornell University Press, 1992); Lizabeth Cohen, *Making a New Deal: Industrial Workers in Chicago, 1919–1939* (New York: Cambridge University Press, 2008); Donald B. Cole, *Immigrant City: Lawrence, Massachusetts, 1845–1921* (Chapel Hill: University of North Carolina Press, 2002); John T. Cumbler, *Working-Class Community in Industrial America: Work, Leisure, and Struggle in Two Industrial Cities, 1880–1930* (Westport, CT: Greenwood Press, 1979); Alan Dawley, *Class and Community: The Industrial Revolution in Lynn* (Cambridge, MA: Harvard University Press, 1976); Thomas Dublin, *Women at Work: The Transformation of Work and Community in Lowell, Massachusetts, 1826–1860* (New York: Columbia University Press, 1981); Paul Faler, *Mechanics and Manufacturers in the Early Industrial Revolution: Lynn, Massachusetts, 1780–1860* (Albany: State University

of New York Press, 1981); Michael Frisch, *Town into City: Springfield, Massachusetts, and the Meaning of Community, 1840–1880* (Cambridge: Harvard University Press, 1972); Laurence F. Gross, *The Course of Industrial Decline: The Boott Cotton Mills of Lowell, Massachusetts, 1835–1955* (Baltimore: Johns Hopkins University Press, 1993); Tamara K. Hareven and Randolph Langenbach, *Amoskeag: Life and Work in an American Factory-City* (New York: Pantheon Books, 1978); William F. Hartford, *Working People of Holyoke: Class and Ethnicity in a Massachusetts Mill Town, 1850–1960* (New Brunswick, NJ: Rutgers University Press, 1990); William F. Hartford, *Where Is Our Responsibility? Unions and Economic Change in the New England Textile Industry, 1870–1960* (Amherst: University of Massachusetts Press, 1996); Paul E. Johnson, *A Shopkeeper's Millennium: Society and Revivals in Rochester, New York, 1815–1837* (New York: Hill and Wang, 1978); Kathy Peiss, *Cheap Amusements: Working Women and Leisure in Turn-of-the-Century New York* (Philadelphia: Temple University Press, 1986); Jonathan Prude, *The Coming of Industrial Order: Town and Factory Life in Rural Massachusetts, 1810–1860* (Cambridge: Cambridge University Press, 1983); Roy Rosenzweig, *Eight Hours for What We Will: Workers and Leisure in an Industrial City, 1870–1920* (Cambridge: Cambridge University Press, 1983); Mary P. Ryan, *Cradle of the Middle Class: The Family in Oneida County, New York, 1790–1865* (New York: Cambridge University Press, 1983); Christine Stansell, *City of Women: Sex and Class in New York, 1789–1860* (Urbana: University of Illinois Press, 1987).

2. Hartford, *Where Is Our Responsibility?*; Cumbler, *Working-Class Community in Industrial America*, 118–20.

3. Hartford argues that the Fall River model of unionization was one "in which a craft elite established standards that other groups then adopted as they formed their own organizations." Hartford, *Where Is Our Responsibility?*, 19, 34.

4. Massachusetts Child Labor Committee, *Child Labor in Massachusetts: Annual Report of the Massachusetts Child Labor Committee, January 1, 1912* (Boston: Massachusetts State Child Labor Committee, 1912), 5, 31.

5. Hine photograph 2630.

6. Hine photograph 2395.

7. U.S. Bureau of Labor, *Report on Condition of Woman and Child Wage-Earners in the United States*, vol. 1 (Washington, DC: Government Printing Office, 1910), 1:217.

8. Ibid.

9. Hine photograph 4216.

10. Hine photograph 4212.

11. National Child Labor Committee, *Child Labor Bulletin* 6, no. 1 (New York: National Child Labor Committee, 1917): 36.

12. U.S. House of Representatives, *The Strike at Lawrence, Mass.: Hearings before the Committee on Rules of the House of Representatives on House Resolutions 409 and 433, March 2–7, 1912* (Washington, DC: Government Printing Office, 1912), 169–73. See also David Brody,

*Workers in Industrial America: Essays on the Twentieth Century Struggle* (New York: Oxford University Press, 1980).

13. "Wants Children Kept Off Dumps," *Boston Daily Globe,* January 1, 1915.

14. Massachusetts Child Labor Committee, *Child Scavengers: Report of the Massachusetts Child Labor Committee, January 1, 1915* (Boston: Massachusetts State Child Labor Committee, 1915), 1.

15. Ibid., 3.

16. Ibid., 3–4.

17. Friedrich Engels, *The Condition of the Working-Class in England in 1844,* trans. Florence K. Wischnewetzky (New York: Cosimo, 2008), 115–16.

18. The identity of Mr. Tebbutt is uncertain, but Mrs. Gertrude Tebbutt was head worker at the King Philip Settlement House in Fall River. *Fall River Directory 1915,* no. 45 (Boston: Sampson and Murdock, 1914).

19. Massachusetts Bureau of Statistics of Labor, *Labor Bulletin of the Commonwealth of Massachusetts, 1906,* nos. 39 to 44 (Boston: Wright and Potter Printing, 1906), 179, 192; John Golden, "The Uprising of the Textile Workers," *American Federationist* 19, no. 9 (September 1912): 730–31; Hartford, *Where Is Our Responsibility?,* 28. See also Blewett, *Constant Turmoil;* Blewett, *The Last Generation;* Cameron, *Radicals of the Worst Sort;* Cumbler, *Working-Class Community in Industrial America;* Dawley, *Class and Community;* Hartford, *Working People of Holyoke;* Rosenzweig, *Eight Hours for What We Will.*

20. Hine photograph 2230.

21. Justus Ebert, *The Trial of a New Society* (Cleveland: I. W. W. Publishing Bureau, 1913), 76–77. Police in Manchester, New Hampshire, turned back a contingent, refusing to allow them to stop in the city.

22. Giovanni DiGregorio and Oscar Mazzitelli, "The Children of Lawrence in New York," *Il Proletario,* February 16, 1912.

23. Diabeti Massimo, "The Exiles from Lawrence to New York," *Il Proletario,* February 16, 1912.

24. Charles P. Neill, *Report on Strike of Textile Workers in Lawrence, Mass. in 1912* (Washington, DC: Government Printing Office, 1912), 50–52.

25. Ibid., 59.

26. Ibid., 504.

27. Ibid., 57.

28. Ibid., 57–58.

29. Ebert, *The Trial of a New Society,* 60–61.

### 10. TRADES AND VOCATIONAL EDUCATION

1. Hine photograph 4660; see also photographs 4661 and 4662.

2. "Points to Its Benefits," *Boston Daily Globe,* January 25, 1909.

3. Ibid.

4. F. Spencer Baldwin, "Recent Massachusetts Labor Legislation," *Annals of the American Academy of Political and Social Science* 33, no. 2 (March 1909): 73.

5. "Points to Its Benefits."

6. *Report of the Commission on Industrial and Technical Education, Submitted May 24, 1905* (Boston: Wright and Potter Printing, April 1906), 1; Baldwin, "Recent Massachusetts Labor Legislation," 63–76.

7. *Report of the Commission on Industrial and Technical Education*, 2–3.

8. Baldwin, "Recent Massachusetts Labor Legislation," 72–73.

9. *Report of the Commission on Industrial and Technical Education*, 5–6.

10. Al Priddy, *Out to Win* (Boston: Massachusetts Child Labor Committee, 1917), 9.

11. E. O. Holland, "Child Labor and Vocational Work in the Public Schools," *Child Labor Bulletin* 1, no. 1 (1912): 16–17.

12. *Report of the Commission on Industrial and Technical Education*, 16–17; U.S. Department of Labor, Bureau of Labor Statistics, *Industrial Experience of Trade-School Girls in Massachusetts*, Bulletin no. 215 (Washington, DC: Government Printing Office, 1917).

13. *Report of the Commission on Industrial and Technical Education*, 16–17; Charles H. Winslow, *Report on the Relations of European Industrial Schools to Labor*, Massachusetts Commission on Industrial Education, Bulletin no. 10 (Boston: Wright and Potter Printing, 1908), 1–22.

14. Holland, "Child Labor and Vocational Work in the Public Schools," 17–23; Georg Kerschensteiner, "Industrial Continuation Schools of Munich," Appendix C, *Report of the Commission on Industrial Education, Submitted June 26, 1906*, Public document no. 76 (Boston: Wright and Potter Printing, 1907), 46–51; Kerschensteiner, *Education for Citizenship* (New York: Rand McNally, 1911), 25; Massachusetts Commission on Industrial Education, *The Agricultural School*, Bulletin no. 7 (Boston: Wright and Potter Printing, 1907), 32–57; Winslow, *Report on the Relations of European Industrial Schools to Labor*, 1–22.

15. Holland, "Child Labor and Vocational Work in the Public Schools," 20.

16. David Montgomery, *Workers' Control in America: Studies in the History of Work, Technology, and Labor Struggles* (Cambridge: Cambridge University Press, 1979); Cecelia Bucki, "Dilution and Craft Tradition: Munitions Workers in Bridgeport, Connecticut, 1915–19," in *The New England Working Class and the New Labor History*, ed. Herbert G. Gutman and Donald H. Bell (Urbana: University of Illinois Press, 1987).

17. To view some of the work Hine did for the American Red Cross, see Daile Kaplan, *Photo Story: Selected Letters and Photographs of Lewis W. Hine* (Washington, DC: Smithsonian Institution Press, 1992).

18. See also Hine photograph 4699.

19. Robert Macieski, "The Home of the Workingman Is the Balance Wheel of Democracy," *Journal of Urban History* 26, no. 6 (September 2000): 715–39.

20. Massachusetts Commission on Industrial Education, *Report on the Advisability of*

*Establishing One or More Technical Schools or Industrial Colleges,* Bulletin no. 11 (Boston: Wright and Potter Printing, 1908), 3–4.

21. Priddy, *Out to Win,* 9.

22. Mary Van Kleeck, Committee on Women's Work of the Russell Sage Foundation, *Working Girls in Evening Schools: A Statistical Study* (New York: Survey Associates, 1914), 168.

23. Margaret Dreier Robins, in *Life and Labor,* August 1913, 231, quoted in Van Kleeck, *Working Girls in Evening Schools.*

24. David Snedden, *The Problem of Vocational Education,* Riverside Educational Monographs (Boston: Houghton Mifflin, 1910).

25. Benjamin C. Gruenberg, "Some Economic Obstacles to Educational Progress," *American Teacher* 1 (September 1912), 90, quoted in Raymond E. Callahan, *Education and the Cult of Efficiency* (Chicago: University of Chicago Press, 1962), 121.

26. John Bellamy Foster, "Education and the Structural Crisis of Capital: The U.S. Case," *Monthly Review* 63, no. 3 (July–August 2011): 6–37.

# Index

*Page numbers in italics refer to illustrations.*

Abels, Margaret Hutton, 185

Adams, Mass. *See* Berkshire Mills

Addams, Jane, 4, 10, 83, 270n14

Adler, Felix, 4

African Americans, 50, 135, 146, 149

Aldrich, Thomas Bailey, 18

Alpert, Hyman, 44

American Sumatra Tobacco Company, 143–44

American Woolen Company, 59, *60*

Amoskeag Manufacturing Company
(Manchester, N.H.), 66–69, 71–74

Anderson, Mary, 185

Anthony, R.I., 50, 51. *See also* Quidwick
Company Mill

apparel industry. *See* garment industry

Appleton, William Sumner, 17

apprenticeship programs, 60, 246–47, 250, 253

Arkwright, R.I., 50, 51. *See also* Interlaken Mill

Arts and Crafts movement, 17

Attleboro, Mass., 196

"authenticity," 6, 77, 162, 170

Baldwin, F. Spencer, 243

Bancroft Foote Boys' Club (New Haven), 44

Barre, Vt., 232

Barthes, Roland, 10–11, 13

Bates Manufacturing Company (Lewiston,
Me.), 56–57, *58*, 59

Beard, Mary Ritter, 42

Bennington, Vt., *1*, 130–31

Benson, Fred, 121–22

Benson, Susan Porter, 179

Berger, John, 5

Berkshire Mills (Adams, Mass.), 208–9

*Bitter Cry of the Children, The* (Spargo), 34,
104–5

Blair, Bertha, 185, 199

Bodeon, Jo, *61, 62, 63*

Bonanno Laundry (Boston), 254

Boscawen Manufacturing Company (Penacook, N.H.), 65–66

Boston, Mass., 16, 133, 177–78, 184, 236; homework in, 184, 186, 193 (*see also under* Roxbury); scavenging in, 223, 224; schooling
in, 93–98, 99; street trades in, 25, 38–41; and
vocational education, 242, 244, 247, 250–51,
254, 256. *See also* Boston-1915 movement;
"1915" Boston Exposition

Boston Industrial Development Board, 177–78

Boston-1915 movement, 80–84, 99. *See also*
"1915" Boston Exposition

Boston School Newsboys' Association, 38–40

Boston Trade School for Girls, 247

Bowman, Fannie, 236–38

Bravas. *See* Cape Verdeans

"Bread and Roses" strike (Lawrence, Mass.,
1912), 221, 229–34, 263

Bridgeport, Conn., 16, 144, *145*, 232; homework

Bridgeport, Conn. (*continued*)
in, 185, 187, 196, 198, 203–4; street trades in, 22, 24, 25, 27, 28
Britain, 13, 186
Brockton, Mass., 244
Brown, Carl, 128–29
Brown, Dona, 17
Brown, Julie K., 78
Brown, O. H., 194
Brownstein, Bessie, 32–34, 35
Buffalo, N.Y., 38, 154, 157
Burlington, Vt., 1, 42. *See also* Chace Cotton Mill
Burnham, Daniel, 81
Byrne, Harriet A., 185, 199

Cambridge, Mass., 16, 183, 186, 256–57, 258
canning industries, 104. *See also* sardine canning
Cape Cod, 131, 133–34
Cape Verdeans, 133–36
Card, Addie, *xiv*, 1–4, 13, 72
Caruso family (Providence), 164
Carver, Mass., 132
Casale, Tony, 31–32, 33
Cella family (South Framingham, Mass.), 199, 200
Central Falls, R.I., 16, 161, 172
Chace Cotton Mill (Burlington, Vt.), 59, 60i 61, 62
*Charities and the Commons* journal, 7, 32
Charity Organization Society, 10
Chase, William Merritt, 16
Chasse, Charles, 64–65
Chicago, 100–101, 154; 1893 World's Fair in, 78, 81
Chicago Plan, 81
Chicago School of Civics Social Museum, 100–101
Chicopee, Mass., 16, 186, 201, 218, 219
*Child Labor Bulletin*, 249
child labor laws, 103–4, 122, 151, 182–83; in Connecticut, 144, 145; and homework, 187, 203; in Maine, 103–6; in Massachusetts, 177–78, 214, 219, 235; regional differences in, 4, 48, 49; in Rhode Island, 53–55, 156, 158
child labor movement, 3, 11, 14, 263. *See also* National Child Labor Committee
child welfare exhibits, 153–57, 175–78. *See also*

Rhode Island Child Welfare Conference and Exhibit
Christmas, Mary, 137
cigarettes, 56, 59, 167–68, 169, 216
Cigar Makers' International Union, 166, 183, 184
cigar making, 165–66, 183–84
Clark, Eugene, 118
Clark, Lotta A., 83
Clopper, Edward, 23–24, 36–37
Cobb, William T., 105
Cocheco Manufacturing Company (Dover, N.H.), 62–63, 64
Cohen, Irene, 158, 159
Cole, John N., 177–78
Cole, Thomas, 17
College Settlements Association, 185
*Collier's* magazine, 7–8
Columbia University, 5, 49
commercial entertainments, 226–28
compulsory education laws, 16, 40, 50, 105, 123, 214; in Massachusetts, 93–94, 235; regional differences in, 50, 75
Conant, Richard K., 131
Connecticut, 18, 184; child labor laws in, 144, 145; street trades in, 9, 22–23, 24–38, 42–47, 271n21; tobacco farming in, 143–52. *See also specific cities*
Connecticut Labor Bureau, 144
Connecticut Leaf Tobacco Association, 149
Connecticut River Valley. *See* tobacco farming
Consumers' League of Connecticut, 150
continuation schools, 247, 248, 249, 254 cranberry farming, 123; housing in, 124, 141; immigrants and, 133–36, 141; in Massachusetts, 123–24, 131–42, 277–78n19
Cromwell, Conn., 146–47
Crossley, George, 71–72
Cumbler, John T., 208
Cushing, Grafton D., 178
Cybalski Tobacco Farm (Hazardville, Conn.), 147

Davis, Philip, 24–25, 44, 46–47
De Farsee (or De Farzen), Jennie, 198, 199
Delloiacono family (Providence), 165–66
De Lucco, Joseph, 32, 33
De Natale, Frank, 256, 257
deskilling, 249–50
Devine, Edward, 10, 11–13

Dewey, John, 4, 260–61
DiGregorio, Giovanni, 232–33
Dillingham Commission on Immigration, 135–36
Dimock, George, 3–4
disease, 22, 162, 180, 198. *See also* tuberculosis
Dismorr, Margaret S., 185
Donovan family (Roxbury, Mass.), 188, *189*
Dorr, Rheta Childe, 182–83
Douglas, William Lewis, 228, 243
Douglas Commission. *See* Massachusetts Commission on Industrial and Technical Education
Dover, N.H., 16. *See also* Cocheco Manufacturing Company
dressmaking, 96–97, *98*, 247
Drown, Frank S., 185
Du Bois, W. E. B., 78, *79*
dumps, 88–89, *90*, 223–24
Duncan, James, 83

Easthampton, Mass., 186
East Hartford, Conn., 151
Eastman, Crystal, 151
Eastport, Me., 104, 106–7, 111, 119, *120*, 121–22. *See also* Seacoast Canning Company
East Wareham, Mass., 133
Eclipse Mills (North Adams, Mass.), 210, *211*, 216–17
Ellis Island, 5–6
empathy, 111, 128; as a goal for Hine, 9, 19, 57, 132
Engels, Friedrich, 226
Ettor, Joseph, 234
Europe, 100, 101–2, 154, 231, 248–49; immigrants from, 18, 82, 94–96. *See also* First World War; *specific nations*
evening classes, 246, 247

Fall River, Mass., 16, 169–70; cranberry pickers from, 133, 134, 141; schooling in, 209, 219, *220*; textile industry in, 213–14, *215*, 239–41, 272n2; unions in, 228–29, 284n3; vocational education in, 244, *245*, 246, 251, *255*, 256, 261, *262*; working-class life in, 206, 207, 208, 210, 213–15, 223–24, 227
family farms, 123, 124–31; in Massachusetts, 124–28; in Vermont, 128–29, 130–31
farmwork, 123. *See also* cranberry farming; family farms; tobacco farming

Fartado, Mary, 51, *52*
Fedele, Annie, 200, *201*
federal government, 74–76, 106, 260. *See also specific agencies*
Fernande, Charlie, 138
Filene, Edward Albert, 81
Finns, 133
First World War, 146, 152, 263; labor shortages during, 144, 149, 152, 249, 253
Fiskeville, R.I., 50
Fitchburg, Mass., 244
Fitzgerald, John Francis "Honey Fitz," 82
Forsythe, Grayson, 108, *110*
Foss, Eugene, 233
Foster, Morrison, 41
Fournier (or Fourner), Henry, 211–12
France, 101–2, 186–87
French Canadians, 2, 69, 83, 133
Froebel, Friedrich, 10

garment industry, 184; and homework, 183, 184, 186, 192–94, 201, 282n46
Gay family (Attleboro, Mass.), 196
gender, 211–13, 281n9; and agriculture, 128, 146–47; and newspaper selling, 29, 32–33; and vocational education, 97–97, *98*
General Federation of Women's Clubs, 21
George family (Worcester, Mass.), 197, *198*
Georgia, 16, 48, 72, 74–75
Germany, 81, 101, 186, 248–49
Gibbons, James, 190, *195*
Golden, John, 243
Goldman, Alice, 32–34, *35*
Goldman, Besie, 32–34, *35*
Goldthwait, Joel E., 224
Goodall Worsted Company (Sanford, Me.), 55–56
Goodell family (Eastport, Me.), 108, *110*, 121
Goodrich Farm (Cromwell, Conn.), 146–47
Gordon, Linda, 183
Greater Berlin movement, 81
Great Falls Manufacturing Company (Somersworth, N.H.), 64–65
Gruenberg, Benjamin, 260
Guimond, James, 10, 16

Hall, G. Stanley, 14
Hamilton, Alexander, 13
Hamilton, Alice, 151

Hamilton family (Eastport, Me.), 119–21

Harrison, Shelby Millard, 78

Hartford, William F., 208

Hartford, Conn., 16, 144, 149, *150*; street trades in, 22, 25–34, 36, *37*, 42–44

Harvard Social Museum, 77, 99–102

Hazardville, Conn., 147

health hazards, 150–51, 159; and homework, 197–203. *See also* injuries

"helpers," 64

Hewes, Amy, 185, 187

Hill Manufacturing Company (Lewiston, Me.), 57, *58*

Hine, Douglas Hull, 4

Hine, Sarah Hayes, 4

Hollow Brook Bog, 136, 139

homework, 163–65, 179–80; defined, 179; effect of, on schooling, 197; efforts to regulate, 179, 183–84, 200, 201–3; in Europe, 186–87; health hazards in, 196–201; Hine's methods of photographing, 165, 187–90; investigations of, 184–87, 197, 199; reformers' objections to, 179, 180–82; wages for, 179, 182, 188–89, 190, 192

Horn, Mery, 29–31

housing, 21, 78, 180; in cranberry farming, 124, 141; in Maine canning industry, 121; in Rhode Island cities, 162–74

Hudson, J. Ellery, 54–55

Hull House, 79


illiteracy, 197

*Il Proletario*, 232–33

immigrant children, 146, 153; and homework, 163–68; reformers' hopes and worries for, 18, 59, 96, 99, *100*, 118, 146; in textile industry, 59, 74–75

immigrants, 17–18, 51, 133–36; and Americanization, 51, 82, 84, 94–96, 141; from Canada, 2, 70; in cranberry farming, 133–36, 141; at Ellis Island, 5–6; Pittsburgh Survey and, 5, 6; prejudice against, 18, 21, 22, 50, 59, 74; in Rhode Island cities, 163–75. *See also* immigrant children

Immigration Restriction League, 18

Indian Manufacturing Company (Indian Orchard, Mass.), 214–16

industrial education. *See* vocational education

industrial schools, 157, 243, 245–47. *See also* textile schools

infant mortality, 169–70

injuries, 221, 224; in family farming, 129; Hine's photographs of, *116*, 129, 239–41, *242*; in Maine canneries, 108, 111, 114–15, *116*

Interlaken Mill (Arkwright, R.I.), 50, *53*

International Association of Machinists (IAM), 253

interpretation, 10, 26, 102; captions and, 114, 124, 126, 266n15. *See also* meaning

Ipswich Mills (Ipswich, Mass.), *247*, *248*

Italians, 163–69, 232–33

Italian Socialist Federation, 232


Jacobson, Matthew Frye, 134–35

Johnson, George E., 82

Jones, Mary Graham, 42–44


Kane, Delia, 257–58, *259*

Kansas City, Mo., 154, *155*, 156

Käsebier, Gertrude, 16–17

Katz, Harry, 236, *237*

Kelley, Florence, 4, 7, 16, 82–83, 181, 226; on children in street trades, 37, 38, 40–41

Kellogg, Paul U., 6, 7, 9, 82, 83

Kerr Thread Company (Fall River, Mass.), 219, 220–22

Kerschensteiner, Georg, 249

Kingsbury, Susan M., 243–44


La Barge, Mamie, 212–13

lace industry, 185, 187, 197, 199

Laird, Addie. *See* Card, Addie

Landwirth, Henry, 232

Lawrence, Mass., 16, 88, 206, *207*, 244; textile industry in, 221, 229–32 (*see also* "Bread and Roses" strike)

Lee, William J., 226

Leeds, Mass., 186, 198–99

Lewiston, Me., 55, 56–59

Lindsay, Samuel McCune, 49

Londry, Alphonse, 133, *134*

Lonsdale, R.I., 158

Lord, E. W., 67, 74

Lorraine Manufacturing Company (Westerly, R.I.), 53

Louisville, Ky., 154, 156

Lovejoy, Owen R., 22–23, 82

Lowell House Social Settlement (New Haven), 44

Lubec, Me., 104, 106–7, 121–22

Lynn, Mass., 244

Macfarland, Henry B. F., 82
Maderyos, Carrie, 138
Maine, 17, 18; child labor laws in, 103–6; textile industry in, 55–59. *See also* sardine canning
Manchester, N.H., 16, 69–70, 166, 169–70, 273n23, 285n21. *See also* Amoskeag Manufacturing Company
Manning, Joe, 2–3
Manny, Frank A., 4
Marx, Karl, 13, 280–81n8
Massachusetts: child labor laws in, 177–78, 214, 219, 235; compulsory education laws in, 93–94, 235; cranberry farming in, 123–24, 131–42, 277–78n19; textile industry in, 197, 211–13, 228–30, 233–34, 246 (*see also under* Fall River); vocational education in, 242–48, 249, 250–59, 261. *See also specific cities and agencies*; Massachusetts Child Labor Committee
Massachusetts Board of Labor and Industries, 201–3, 219
Massachusetts Bureau of Statistics, 179, 185, 190
Massachusetts Charitable Mechanics Association, 247
Massachusetts Child Labor Committee, 178, 180, 219, 249; Hine's work with, 131, 211, 212, 250; on scavenging, 223, 224; on vocational education, 249, 250, 256, 261
Massachusetts Commission on Industrial and Technical Education, 243–45, 246, 248, 254–56
Massachusetts Minimum Wage Commission, 185
Massimo, Diabeti, 232
Mazzitelli, Oscar, 232–33
McIntire, Ruth, 123
McKelway, A. J., 48–49
meaning, 8, 9, 10–11, 15, 26; captions and, 24, 37, 114, 266n15
mechanization, 13, 107; reformers' hopes for, 106, 122, 148–49, 151
medical photographs, Hine's, 236–39
messenger boys, 23, 34–38, *39*, 148, 271n21
Messina family (Cambridge, Mass.). 256–57, *258*
Mesuretta, Molla, 240–41
midwives, 169–72
Mintz, Steven, 9
Morini family (South Framingham, Mass.), 187–88

mule spinning, 60–62, 158, 256

Natick, R.I., 50
National Child Labor Committee (NCLC), 4, 19–20, 55, 87, 250; end of Hine's work for, 143, 151, 235, 250, 263; and farmwork, 123–24, 131–32, 133, 152; Hine as employee of, 1, 4, 14, 18, 22–23, 42, 53, 106–7, 131, 157, 176, 214, 261–63; publications by, 179, 250; and schooling, 256; and southern textile industry, 176–77; women activists and, 4, 10, 21. *See also* Clopper, Edward; Conant, Richard K., Lord, E. W.; Lovejoy, Owen R.; McIntyre, Ruth; McKelway, A. J.; Smith, F. A.
National Conference of Charities and Correction, 7–9, 155
National Consumers League, 21, 184. *See also* Kelley, Florence
National Society for the Promotion of Industrial Education (NSPIE), 260
National Women's Trade Union League, 21, 259
Nearing, Scott, 32
Neill, Charles P., 34, 76
New Bedford, Mass., 16, 88, 244, 246; cranberry pickers from, 134, 141; working-class life in, 206, 207, 217–18
*New Boston* journal, 82
New Britain, Conn., 154
New Hampshire, 17, 18; textile industry in, 62–72, 220–22. *See also* Dover; Manchester; Portsmouth
New Hampshire Spinning Mills (Penacook, N.H.), 65–66
New Haven, Conn., 16, 22, 37, 44, *45*, 46, 144
New Jersey, 123
Newsboys Building (Boston), 39–40
Newsboys Reading Room (Boston), 84–85
Newsboys' Club (Boston ), 84, 85
newsies (newsboys and newsgirls), 123–24, 156; gender differences among, 29, 32–33; Hine's photographs of, 24–34, 39–41, 43–44, 84–85, 158, *159*; in popular culture, 23; reformers and, 23–25, 32, 38–41, 43–44, 158–59
New York City, 78, 149, 154, 232; Hine and, 181, 184, 185, 265n2; homework in, 181, 184, 185; street trades in, 34, 38
New York State, 78, 124, 129, 184. *See also* Buffalo; New York City
New York University (NYU), 4–5

Nichols, J. Howard, 49
night schools, 248
"1915" Boston Exposition, 80, 82, 274n12; Hine's photographs in, 84–99, *100*, 275nn21,26
Nissenbaum, Stephen, 17
Normandin, Fred, 71
North, Francis R., 157
North Adams, Mass., 206, 210
Northampton, Mass., 154, 155, 186, 201
North Bennet Street Industrial School, 247
North Carolina, 176–77
North End Union (Boston), 247
North Pownal, Vt., 1, 2. *See also* Card, Addie
Nutting, Wallace, 17

Orvell, Miles, 10–11
*Outlook*, the, 104

Pagnette, Adrienne, 213
Parent, John, 209–10
Parker, Lewis, 50
Parker, Louis W., 83–84
Parker Manufacturing Company (Warren, R.I.), 54
Pawtucket, R.I., 16, 88, 161, 162; textile industry in, 48, 158, *159, 160*
Pawtuxet Valley, 50–51
Peabody, Francis Greenwood, 77, 83
Peiss, Kathy, 31
Penacook, N.H., 65–66
Peter, Gordon, 138–39
Phoenix, R.I., 50
photo-secessionists, 216
Pittsburgh Survey, 5–7, 78, 79; Hine and, 5–6, 76, 77, 100, 275n26; as model for social surveys, 5–6, 77, 78
Pittsfield, Mass., 244
play, 43, 44, 84–85, 99, *100*, 196, 226
Playground and Recreation Association of America, 157
Pocasset Mill (Fall River, Mass.), *245, 246*
Poiriere, Estelle, 239–40
portraiture, 19
Portsmouth, N.H., 16, 18
Portuguese immigrants, 51, *52*, 133–36, 142, 215i. *See also* Cape Verdeans
pragmatism, 11
Priddy, Al, 250
privies, 172–74

Proctor, Vt., 1
professionals: as an emerging social class, 79, 82, 83, 175–76; and expanded role for the state, 185, 236, 261–63; Hine's contacts among, 6–7, 22, 261–63; key role of, in Progressive reform, 4, 21–22, 79, 153–54, 185
Progressive reformers, 3, 4, 15; and education, 4, 10, 96, 220, 242, 249, 260; and public exhibits, 77, 153, 157, 178; faith of, in government actions, 106, 122, 183; Hine and, 3, 6–8, 11–13, 14, 22, 77, 96, 206, 261; middle-class identity of, 11, 14, 22; professionals as, 4, 21–22, 79, 153–54, 185; and social surveys, 6–7, 78–79, 183; women as, 21–22, 96, 185. *See also* National Child Labor Committee
Prosser, Charles, 259–60
Providence, R.I., 16, 161, 192; cranberry pickers from, 133, 134, 141; homework in, 163–66, 192; housing in, 162–64, 165–66, 168–74; scavenging in, 88; street trades in, 34–36, 158, *159*. *See also* Rhode Island Child Welfare Conference and Exhibit
public exhibits, 77–79, 100–101, 153; in Boston, 82, 177–78 (*see also* "1915" Boston Exposition). *See also* child welfare exhibits
Pulk, Hiram, 108–11

Quidwick Company Mill (Anthony, R.I.), 51, *52*
Quintin, Rhea, *255, 256*

race, 50, 78, 134–35, 146; Cape Verdeans and, 134–36
Radcliffe, Jane E., 103
"reality effect," 13
Red Cross, 143, 151, 250, 263
Reed, Warren A., 243
Revinsky, Oscar, 223–24
Rhode Island, 18, 51, 158; child labor laws in, 53–55, 156, 158; homework in, 185, 187, 197, 199; textile industry in, 50–55, 158. *See also* Newport; Pawtucket; Providence; Rhode Island Child Welfare Conference and Exhibit
Rhode Island Child Welfare Conference and Exhibit, 154, 156–57, 174–75; goals of, 157–62; and housing conditions, 162–65, 168–74; and immigrant working class, 165–69
Riis, Jacob, 78–79
River Point, R.I., 50, 51. *See also* Royal Mill
Rosenblum, Walter, 5

Routzahn, Evart Grant, 175–76

Routzahn, Mary Swain, 175–76

Roxbury (Boston neighborhood): homework in, 188, *189*, 190, *191*, 192, 193, 194, *195*

Royal Mill (River Point, R.I.), 51, *52*

Russell Sage Foundation, 6, 78, 176, 182, 258, 275n26. *See also* Van Kleeck, Mary

Saint-Gaudens, Augustus, 17

Salem, Mass., 16, 186, 201, 209–10, 211–12

Sampsell-Willmann, Kate, 11, 14

Sanford, Me., 16, 55. *See also* Goodall Worsted Company

Sanger, Margaret, 232

sardine canning, 103–22, 132–33; and child labor laws, 103–6; injuries in, 108, 111, 114–15, *116*; and schooling, 105, 121–22; wages for, 107–8, 118–19

Sargent, John Singer, 16

Saunders, Lucy, 128

Saunders, Robert, *140*

Scanlon, M. F., 233

scavenging, 86, 87–92, 223–26

school attendance, 96, 197, 208, 235–36, 242–43; agricultural work and, 129, 142, 151; in Maine, 105, 121–22; parents' attitudes toward, 69–70. *See also* compulsory education laws

scientific management, 259

Seacoast Canning Company, 108–17, *120*, *121*, 122

Sekula, Allan, 19

Sewall, John L., 82

sexism, 217–18

Shaw, Elsie, 117, *118*

Shorey, Eva L., 105–6

Silverberg, Helene, 21

Slater, Samuel, 48

Smith, F. A., 186

Smith, Hoke, 260

Smith, Judith E., 163–64

Smith, Shawn Michelle, 78

Smith-Hughes Act (1917), 260–61

Snedden, David, 246

Socialist Party of America, 7, 232

social museums, 99–102. *See also* Harvard Social Museum

social photography, 7–9

social surveys, 78–79, 123–24, 185; Hine and, 42, 77, 131–32, 143–45, 186, 230, 265n2; of homework, 184–87, 188–89, 190–92, 197–98, 203,

204; importance of, to reformers, 6–7, 78–79, 184; Pittsburgh Survey as model for, 5–6, 77, 78

social work, 5–7, 8, 9, 11–13, 185, 189–90

Society for the Preservation of New England Antiquities, 17

Society for the Protection and Care of Children, 223

Somersworth, N.H. *See* Great Falls Manufacturing Company

Somerville, Mass., 16; homework in, 186, 194, 198, *199*, 200–201, *202*

Sousa, Amelia Louise, *143*

South, the, 4, 146. *See also* southern textile industry

South Carolina, 16, 48, 50, 56

South Carver, Mass. *See* T. B. Smart Bog

South End Settlement House (Boston), 242

southern textile industry, 48–50, 60, 176–77, 214; comparisons of New England mills with, 15–16, 20, 49, 56–57, 74–75, 76, 214; Hine and, 51, 56–57, 72

South Framingham, Mass., 187–88, 199

South Windsor, Conn. *See* American Sumatra Tobacco Company

Spargo, John, 34

spinning frames, *xiv*, 1–2, 51, *53*, 72, *73*, 211–13

spinning processes, 60–62. *See also* mule spinning, 158, 256

Sprague House Settlement (Providence), 174–75

Springfield, Mass., 16, 186, 201, 244

Stange, Maren, 7

state, the, 24–25, 42–47; expanded role of, 183, 185, 235, 236, 261, 281n24; reformers' faith in, 106, 122, 183

Stewart, Clinton, 129

Stieglitz, Alfred, 16–17

strikes, 184, 228–31. *See also* "Bread and Roses" strike

Suncook, N.H., 66. *See also* Suncook Mills

Suncook Mills (Suncook, N.H.), 66, *67*

Sweetser, E. Le Roy, 232

Syrians, 133

Tagg, John, 190

Taylor, Harry C., 250–51

Taylorism, 259

T. B. Smart Bog (South Carver, Mass.), 133, *134*, 137, *140*

Tenement-House Exhibition of 1899, 78

Teoli, Camella, 221

textile industry, 42, 48, 256; in Maine, 55–59; in Massachusetts, 197, 211–13, 228–30, 233–34, 246 (*see also under* Fall River); in New Hampshire, 62–72, 220–22; in Rhode Island, 50–55, 158; in the South, *see* southern textile industry; in Vermont, 59–62 (*see also* Card, Addie); wages in, 208, 228, 231, 241–42

textile schools, 246, 256

theft, 225–26

Thomas, Minnie, 111–14

Thomas, Phoebe, 114–15, *116*

Thwing, Charles F., 83

tobacco farming, 143–52; wages in, 146, 147–48

Trachtenberg, Alan, 15, 42

trade schools, 245–46, 247, 250

truancy, 93–95

tuberculosis, 32, 162, 175, 198, 199

union labels, 166, 184

United Garment Workers of America, 183

United Textile Workers Union, 243

United Workers Boys' Club (New Haven), 44–46

University of Chicago, 4, 261

U.S. Bureau of Labor, 214

U.S. census, 3, 13–14, 104, 123, 164, 166

U.S. Children's Bureau, 154, 185

U.S. Commission of Fish and Fisheries, 106

U.S. Department of Agriculture, 141

U.S. Department of Commerce and Labor, 76

U.S. Department of Labor, 185. *See also* U.S. Children's Bureau; U.S. Women's Bureau

U.S. Postal Service, 2

U.S. Women's Bureau, 185, 186, 203–4

Van Kleeck, Mary, 182

Veiller, Lawrence, 78, 82

Vermont, 1, 18, 42, 128–29, 130–31, 265n1; farmwork in, 128–29, 130–31; textile industry in, 59–62 (*see also* Card, Addie). *See also* Barre; Burlington; West Pownall

Vernon, Conn., 147

Visiting Nurses Association, 170

vocational education, 235, 241–42; conflicting views of, 245–46, 260–61; in Europe, 248–49; gender differences in, 96–97, 98i, 258–59; in Massachusetts, 242–48, 249, 250–59, 261; as new focus for Hine's work, 250–54, 261–63; reformers' hopes for, 10, 96, 220, 242

wages, 70, 137; and homework, 179, 182, 188–89, 190, 192; in sardine canning, 107–8, 118–19; in textile industry, 208, 228, 231, 241–42; in tobacco industry, 146, 147–48; withholding of, 141

Wald, Lillian, 23

Warren, R.I., 53–54

Warren Manufacturing Company (Warren, R.I.), 53–54

Weatogue, Conn., 147, *148*

Weeks family (Williamsburg, Mass.), 190

Wells Memorial Institute, 247

Westerly, R.I., 53

Westerly Mills (Westerly, R.I.), 53

West Springfield, Mass., 186, 201

Wetstone Farm (Vernon, Conn.), 147

Wexler, Laura, 11, 26

Whitin, Ernest Stagg, 104, 106

Wilcox, William A., 53

Williamsburg, Mass., 190

Wilson, Caroline E., 185

Winchendon, Mass., 211–12

Windsor, Conn., 151

Winooski, Vt., 1. *See also* American Woolen Company

Winslow, Mary N., 186, 198

Winthrop, Elizabeth, 2–3

Wischnewetzky, Florence Kelley. *See* Kelley, Florence

Wise, Stephen S., 49

women: in canning industry, 107–8, 118; in machine trades, 250; as reformers, 21–22, 96, 185. *See also* gender

Women's Educational and Industrial Union (WEIU), 70

"woodpickers," 88, *91*

Woodruff, Clinton Rogers, 82

Woods, Robert A., 242–43

Woonsocket, R.I., 16, 161, 172, 197

Worcester, Mass., 16, 186, 197, *198*, 201, 244

work certificates, 53, 214, 235, 246, 256

work hours, 74–75, 145, 150, 158–59, 203

Wright, Carroll D., 243

Yale University, 44

Young Men's Christian Association (YMCA), 247

Zaikener, Carrie, 232